Super Interesting TECHNOLOGY FACTS For Smart Kids

MARCUS MORALES

Copyright © 2025 by Intel-Excellence

All rights reserved. No part of this publication may be copied, stored in a retrieval system, or transmitted in any form or by any means (electronic, mechanical, photocopying, recording, or otherwise) without the prior written permission of the publisher, except for brief quotations used in reviews or educational settings.

This book is a work of nonfiction intended to educate and entertain young readers. All facts have been presented in a simplified, age-appropriate way. Every effort has been made to ensure accuracy at the time of publication.

Author: Marcus Morales

Publisher: Intel-Excellence

First Edition, June 2025

ISBN: 979-8-9921512-4-4

Printed in USA

For permissions, visit: www.intel-excellence.com

Table of Contents

Introduction 1

Chapter 1: Revolutionary Tech That Changed the World 3

   1. The Wheel That Started It All 3

   2. The Book Machine 5

   3. Lights On! The Bright Idea That Lit Up the World 6

   4. Hello, World! The First Phone Call 7

   5. Beep Beep! How Cars Took Over the World 8

   6. The 12 Seconds That Made Us Fly 9

   7. Giant Computers to Pocket Power 10

   8. To the Moon! 12

   9. Internet Explosion 13

   10. Smartphone Magic 14

   11. Amazing Artificial Hearts 15

   12. Smart Speakers, Smarter Answers 17

   13. Print a Toy? Yes You Can! 18

   14. Solar Energy Saves the Day 19

   15. What's Next? 21

Chapter 2: Gadgets and Gizmos 23

   1. What in the World Is a Gadget?! 23

   2. Smartwatches Are Tiny Wrist Computers! 24

   3. Sky Robots You Can Fly! 25

   4. The Magic of Microwaves 27

5. Buttons, Joysticks and Game Power!.................................. 28
6. Say Cheese! A Snapshot Through History 29
7. The Math Magicians in Your Pocket 30
8. The Invisible Connector 31
9. Meet the Dust-Busting Machines 32
10. Tiny Tech That Takes Your Temperature 33
11. Talk to Your Tech: Meet Smart Speakers 34
12. Books, Games & Movies... All in One Gadget 35
13. Wristbands That Cheer You On! 37
14. Power Up Again... and Again! 38
15. The Coolest Gadget Ever - What's Yours?..................... 39

Chapter 3: Robots and Artificial Intelligence 41

1. Robots Come in All Shapes 41
2. Meet the Robots That Make Life Easier 42
3. Robots in Space? Yep! They're Exploring Mars! 43
4. Robots Work in Factories and Farms 45
5. Want a Pet Dinosaur? Robot Pets Are Real! 46
6. Robots Have a Smart Brain! 47
7. Can You Talk to Robots? 48
8. Can a Robot Tell if You're Happy or Sad? Yes! 49
9. Robot Arms and Legs Are Super Strong 51
10. Robots Go to School Like Kids Too - Kind Of! 52
11. Robots Are Movie Stars 53
12. Robots Follow Instructions 55
13. Some Robots Are Tiny Doctors 56

14. Robots Can Build Huge Things 57

15. Future Robots Could Do Anything 58

Chapter 4: The Internet and You 60

1. Did You Know the Internet Connects the Whole World? 60

2. The Internet Zooms Around So Fast 61

3. The Difference Between the Web and the Internet 62

4. Can Messages Really Travel in Seconds? 63

5. Search Engines Know 64

6. Why Streaming Is Popular 64

7. Can You Really Play Games with People Far Away? 65

8. How Does the Internet Help Us Learn? 66

9. Is It Really Like Talking Face to Face? 67

10. Where Do All My Online Files Go? 68

11. Why It Is Important to Stay Safe Online 69

12. You Can Use the Internet Anywhere 69

13. What Is a Digital Footprint and Why Should You Care? 70

14. Internet Cables Run Across the Ocean Floor 71

15. The Internet Keeps Growing Every Second 72

Chapter 5: Video Games and Virtual Worlds 75

1. How Did Video Games Begin? 75

2. Game Consoles Are Mini Computers 76

3. Controllers Are Your Game Hands 77

4. Games Use Math and Art 78

5. Games That Help You Learn 79

6. Welcome to Virtual Worlds 81

7. People Make Games for a Living 82

8. What Is Augmented Reality? 83

9. You Can Team Up Online 84

10. Grown-Ups Play Too 85

11. You Can Make Your Own Game! Here's How 87

12. You Control Your Avatar 88

13. Why Game Ratings Matter 89

14. Games Are Fun on Purpose 90

Chapter 6: Space Tech: From Earth to the Stars 92

1. Rockets Lift Us Into Space 92

2. Satellites Fly Above Our Heads 93

3. Space Stations Are Homes in Orbit 95

4. Telescopes See Farther Than Ever 96

5. Rovers Explore Other Planets 97

6. Astronauts Use High-Tech Suits 98

7. Moon Missions Changed Everything 100

8. Mars Is the Next Big Goal 101

9. Satellites Help Us on Earth 102

10. Robots Help Build in Space 103

11. Space Food Is Special Science 105

12. GPS Works Thanks to Space Tech 106

13. Space Telescopes Discover New Worlds 107

14. Spacecraft Go Beyond Our Solar System 108

15. Kids Like You Might Go to Space Someday 109

Chapter 7: Green Technology 112

1. Solar Panels Turn Sunshine into Power 112
2. Wind Turbines Spin Up Clean Energy 114
3. Electric Cars Don't Need Gas 115
4. Smart Thermostats Save Energy 116
5. LED Lights Use Less Electricity 117
6. Recycling Turns Trash into New Things.................. 118
7. Green Roofs Help Cool Buildings 119
8. Water Filters Keep Rivers Clean 121
9. Solar-Powered Gadgets Are on the Rise................ 122
10. Bicycles Are Super Eco-Friendly............................. 123
11. Composting Gives Food Scraps New Life 124
12. Eco-Houses Are Built to Save Energy 125
13. Ocean Clean-Up Robots Collect Trash 126
14. Plant-Based Plastics Help the Planet 127
15. Kids Can Be Green Inventors Too! 129

Chapter 8: The Future of Technology............................ 131
1. Flying Cars Are Being Tested Today....................... 131
2. Robots Might Deliver Your Pizza 132
3. 3D Printers Can Build Houses 133
4. Holograms Will Make Screens Float in Air............. 135
5. Smart Glasses Could Replace Phones.................... 136
6. Your Clothes Might Charge Your Devices............... 137
7. AI Will Become Super Smart.................................. 138
8. Underwater Robots Will Explore Deep Oceans 140
9. Space Travel May Become Common 141

10. Smart Homes Will Do Everything for You 142

11. Virtual Reality Will Feel More Real 143

12. Self-Driving Cars Will Take You Places 144

13. Food Will Be Grown in Vertical Farms 145

14. Your Toys Might Talk Back 146

15. Kids Will Invent Tomorrow's Tech 147

Introduction

Did you know that more than a million kilometers of internet cables slither across the ocean floor?

Did you know some jackets today harvest body heat to recharge a smartphone?

Twelve-year-old Shubham Banerjee wanted to help his visually impaired grandmother read. When he discovered that commercial Braille printers cost thousands of dollars, he built his own "Braigo" using Lego Mindstorms and spare electronics. His low-cost device opened new worlds of books for his grandmother and grew into a real company making affordable Braille printers for schools everywhere.

That spark of "What if I could…?" is exactly what drives every chapter of this book. **SUPER INTERESTING TECHNOLOGY FACTS FOR SMART KIDS** takes you on a whirlwind tour of inventions and ideas shaping our world, from rockets roaring into orbit to robots patrolling shipwrecks beneath the waves. Along the way, you'll learn why vertical farms grow leafy greens like skyscrapers, how smart homes learn your favorite cozy temperature, and why 3D printers can stack walls into entire houses.

Each bite-sized fact comes with its own boost: a **Tech Word** to power up your vocabulary, a **Fun Fact** to surprise your friends or a **Challenge** that dares you to invent, experiment and explore. Whether your dream is building flying cars, coding friendly robot buddies, or growing gardens in the sky, this book lights the path from "That's cool!" to "I want to build that!"

Whether you dream of designing flying cars, programming friendly robot helpers or growing skyscraper gardens, this book lights the path from "Hmm, interesting…" to "I want to build that!" Packed with real stories, playful explanations and eye-opening facts, it's a must-read for every curious child who's ever whispered, "Someday, I'll invent something amazing." Turn the page and let your next big idea take flight!

Chapter 1: Revolutionary Tech That Changed the World

The Wheel That Started It All

Did you know that the wheel is one of the oldest and most incredible inventions in the whole world? It was created more than 5,500 years ago! That's even older than pyramids and way before your great-great-great-grandparents were born. People invented the wheel to help them do hard jobs more easily, especially moving heavy things from place to place.

Imagine trying to carry a giant stone or a pile of food without any wheels. You'd have to drag it or carry it, which would take forever and wear you out! But with wheels, people could roll things instead. Much easier, right?

Thanks to the wheel, people could build carts, wagons, and eventually bikes, skateboards, roller skates, cars, and trains. Wheels are used in machines. A few of them have little teeth on them and are called gears. Gears help machines turn, spin, and do meaningful work.

The wheel is everywhere! It's on your toy cars, your bike, your luggage and inside clocks and engines. Some wheels are smooth and round, like the ones on scooters. Others are tiny and have bumpy teeth, like the gears in your robot toys.

So, the next time you're riding your scooter, pedaling your bike, or spinning the wheels on your toy truck, give a little cheer for the wheel. It's one simple invention that helped build the modern world!

Tech Word: Gear - A type of wheel with teeth that fits into other wheels to help machines move.

Figure 1 - Gear

The Book Machine

A long, long time ago, there were no printers, no photocopiers, and no way to make a book fast. Every book had to be copied by hand, one page at a time. That meant someone had to sit and carefully write every word with ink and a feather pen. Imagine copying your favorite storybook by hand. It would take forever!

Then, in 1440, everything changed! A man named Johannes Gutenberg invented something amazing called the printing press. It was a big machine that could press ink onto paper and make many copies of a book in less time. This was a huge deal!

Before the printing press, only a few people had books because they were rare and expensive. However, with Gutenberg's invention, books became cheaper and easier to make. More people could read, learn, and enjoy stories. Schools got better, libraries grew bigger, and people could share ideas more quickly.

Today, we still use printing technology, but it's more advanced. We have color printers for photos, laser printers for schoolwork and 3D printers that can print real objects like toys and tools!

Tech Word: Printing Press - A machine that presses ink onto paper to make books and copies faster than writing by hand.

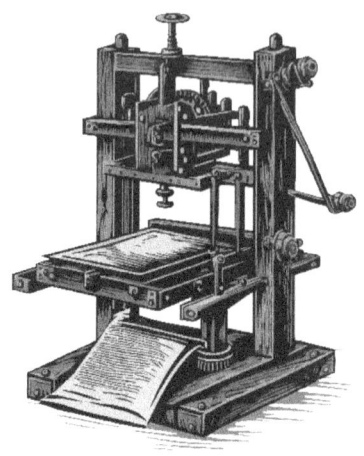

Figure 2 - Gutenberg Printing Press

Lights On! The Bright Idea That Lit Up the World

Did you know Thomas Edison didn't invent the very first light bulb? But he made one that worked much better and lasted a lot longer! Before Edison's light bulb, people used candles and lamps that made rooms dark and smoky. Edison's invention helped light up homes, schools and whole cities all around the world.

His light bulb was special because it used a tiny wire inside called a **filament** that glowed brightly when electricity passed through it. Edison tested thousands of different materials to find the best filament and when he found one that worked, it stayed bright for hours!

This amazing invention changed how people lived. Instead of going to bed when the sun went down, they could stay up and read, play and work. Streetlamps helped keep people safe when it was dark outside.

Edison didn't stop there. He also helped build the first power plants that made electricity that went through wires to homes and businesses. Thanks to Edison and electricity, we have lights, TVs, computers and so much more today.

Tech Word:

Electricity - A form of energy that powers lights, TVs, computers and lots of other cool things!

So next time you flip a light switch, remember Edison helped make that possible!

Hello, World! The First Phone Call

"Mr. Watson, come here, I want to see you." These were the very first words ever spoken on a telephone by Alexander Graham Bell in 1876! It was a huge moment because, before that, people could only talk to each other face-to-face or send letters.

The telephone was a new way for people to talk over long distances. At first, phones were big and heavy with long cords. Over time, phones got smaller and easier to use. Today, we carry smartphones in our pockets that can do so much more than make calls!

Here's a cool fact: your smartphone has a tiny computer inside that is way more powerful than the computers NASA used when they sent astronauts to the Moon in 1969! That means your phone can do amazing things like play games, take pictures and help you learn.

Phones have changed how people stay connected with friends and family, share stories and do their homework. Imagine living in a time without phones; you'd have to wait days or weeks to hear from someone far away!

Thanks to Alexander Graham Bell and all the inventors after him, we can talk to anyone, anywhere, anytime!

Did You Know?

The first words on the telephone were a call for help. What would you say if you were the first person to use a phone?

Beep Beep! How Cars Took Over the World

Cars are everywhere today, but long ago, only a few people could afford them. Henry Ford didn't invent the car, but he made a big change that helped lots of people own one! How? He created the assembly line, a clever way to build cars faster and cheaper.

Before the assembly line, cars were made by hand, one at a time. This took a long time and cost a lot of money. Ford's assembly line was like a super-fast team where each worker did one small job, like putting on a wheel or a door. The car moved down a conveyor belt from one worker to the next until it was finished!

Because cars could be made faster and cost less, many more families could buy them. This changed the world!

People could travel farther, visit friends and explore new places. Cars helped towns grow and made it easier to deliver food and mail.

Today, there are millions of cars zooming around the world. Certain cars run on gas, others use electricity, and scientists are working on cars that can drive all by themselves!

Try This:

Tomorrow, count how many cars you see on your way to school. Are they different colors? Different sizes? What kind of car would you like to drive?

Cars are amazing machines that help move the world every day!

The 12 Seconds That Made Us Fly

Have you ever dreamed of flying like a bird? In 1903, two brothers named Orville and Wilbur Wright made that dream come true. They flew the first powered airplane! Their plane flew for **12 seconds** and went about 120 feet. That might not seem like much, but it was the start of a whole new way to travel!

The Wright brothers built a small, light plane with wings that flapped like a bird. They tested it carefully and finally made it fly with an engine. People were amazed!

Today, airplanes are huge and fly high above the clouds, traveling thousands of miles in a few hours. Planes take people to visit family, go on vacation, or deliver important packages around the world.

Flying wasn't always easy. Early pilots had to be brave because the planes were small and shaky. But thanks to science and smart engineers, modern planes are safer and comfortable.

Tech Word: Aviation is the science of flying planes and helicopters.

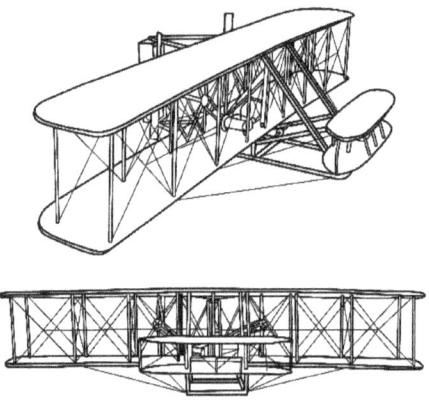

Figure 3 - The Wright Brothers' 1903 Plane

Next time you see a plane in the sky, remember the Wright brothers and their 12-second flight that started it all!

From Giant Computers To Pocket Power

Can you imagine a computer so big it filled an entire room? That was the ENIAC. One of the very first computers ever made! Built in 1945, ENIAC was as long as a school bus and weighed about 27 tons; that's heavier than four elephants! It didn't have a screen or a keyboard. It used over 17,000 vacuum tubes to work and those tubes got very hot.

Even though ENIAC was huge and super slow compared to today's computers, it was an amazing invention at the time. It could solve math problems much faster than a person could with a pencil and paper. People used it to help with science, weather predictions and space research.

Fast forward to today, you have a much more powerful computer in your backpack or maybe in your pocket! Tablets, laptops and smartphones can do so much more than ENIAC ever could. You can play games, watch videos, draw pictures and also talk to people around the world.

Here's something cool: computers keep getting smaller and smarter. A few are so tiny they can fit on the tip of your finger!

Did You Know?

The ENIAC didn't use computer chips like we do today. Instead, it used vacuum tubes that glowed like little light bulbs!

Tech Word: Computer: A machine that stores and processes information. It can help you learn, play, create and connect!

Figure 4 - Modern Desktop Computer

What would YOU use a computer for if you could invent your own?

To the Moon!

On July 20, 1969, something amazing happened. Humans walked on the Moon for the very first time! The mission was called Apollo 11 and it was run by NASA, the United States space agency. Astronauts Neil Armstrong, Michael Collins and Buzz Aldrin rode in a giant rocket all the way to space.

When Neil Armstrong stepped onto the Moon's surface, he said the famous words: "*That's one small step for man, one giant leap for mankind.*"

The world watched in awe as history was made! The astronauts wore special space suits and traveled in a spacecraft that had many parts, including the Apollo 11 Lunar Module, which touched down on the Moon. They planted an American flag, collected Moon rocks and left a tiny plaque that says, "We came in peace for all mankind."

Can you believe this? The Apollo 11 spacecraft had less computer power than a calculator or a modern watch! The astronauts had to be very smart and brave. They practiced for months to make sure the mission would go smoothly.

Now, we use much more powerful technology to explore space. Robots and rovers visit planets, and astronauts live on the International Space Station for months at a time!

Fun Fact: The footprints left by the astronauts are still on the Moon today because there's no wind to blow them away!

What would you do if you could visit the Moon?

Would you bounce in low gravity, take Moon selfies or search for alien life?

Internet Explosion

Can you imagine life without the internet? No videos, no online games, no emailing your teacher or messaging your friends. It's hard to picture, right? But the internet wasn't always around! It all began in the 1960s when scientists wanted their computers to be able to "talk" to each other and share secrets, like a special club!

Over the years, that small secret club grew into something huge. Today, the internet connects billions of people across the world! It's like a giant invisible web that lets you explore websites, watch shows, play games and video call someone on the other side of the planet.

The internet helps with school, too. You can learn about dinosaurs, space or how to do a cool science experiment;

all from your tablet or computer. And guess what? Some people actually use the internet to control robots, fly drones, or watch live videos from space!

And how do we connect to the internet? One magical way is through **Wi-Fi**! Wi-Fi sends invisible signals through the air so your phone, tablet or computer can go online without any wires. It's like your devices are talking through the air!

Tech Word: Wi-Fi - A special way to connect to the internet using invisible radio waves with no cords needed!

So, next time you're watching your favorite cartoon or looking up jokes online, remember the internet makes it possible!

Smartphone Magic

Imagine carrying a tiny computer, a camera, a flashlight, a calculator, a map and a video game console - all in your pocket! That's what a smartphone can do!

The very first smartphone was called the IBM Simon and it came out in 1992. It was big and chunky, and it could do a few things, like send emails and make calls. Today's smartphones are way cooler and much smarter! You can send texts, take amazing photos, play music, record videos, use apps and ask questions using your voice.

Most smartphones have face scanners or fingerprint readers to unlock them, just like in spy movies! Others can help you find your way if you're lost, count your steps, or teach you how to play the piano.

Smartphones are used by teachers, doctors, chefs and astronauts. They're also great for fun stuff. Drawing, listening to stories and chatting with friends or grandparents far away.

Want to know why your phone works so fast? It's because it has a tiny, super-smart brain inside called a microchip!

Challenge: What's your favorite app on a smartphone? Is it a drawing app? A fun game? Or something that helps you learn?

Figure 5 - IBM Simon smartphone

Remember: smartphones are awesome tools, but they're more amazing when you use them to create, learn and explore!

Amazing Artificial Hearts

Your heart is a super-important organ. It works like a pump by sending blood around your body to keep you alive and healthy. But sometimes, hearts get sick and don't work as they should. That's where **technology** steps in to help!

Scientists and doctors have made special **artificial hearts** that are like clever machines placed inside people to help their blood pump. It's like giving someone a high-tech helper to keep them alive!

The first artificial heart was used in 1982 and looked like something from a science fiction movie. It was big, bulky, and powered by a machine outside the body. Today, artificial hearts are smaller, smarter and can fit inside a person's chest, giving them more freedom and time while waiting for a heart transplant.

The people who design these lifesaving machines are part of a field called biomedical engineering. That means they mix biology (the study of the body) with engineering (the design of machines) to create tools that help people feel better.

Tech Word: Biomedical Engineering - A science that combines medicine and machines to help people live healthier lives.

Isn't it amazing how technology can keep a heart beating? Maybe one day, you'll invent a machine that helps people too!

Smart Speakers, Smarter Answers

If you've ever said, "Hey Siri" or "Alexa, what's the weather?" you've talked to a smart helper inside your phone or a speaker! These friendly voices inside your phone, tablet and speaker can do all kinds of things. Play music, tell jokes, answer questions, set timers and turn off the lights!

Smart assistants use something called artificial intelligence, or AI for short. With AI, your computer can THINK and LEARN in a way that's kind of like a human brain. It doesn't have feelings, but it can study patterns and data, like how many times you ask it to play your favorite song or when you usually wake up.

These smart assistants learn over time. If you always ask for jokes, they'll remember and get better at telling funny ones! If you ask for help with spelling, they'll help you sound out tricky words. A few help grown-ups control their homes, like turning on the oven, adjusting the thermostat, or locking the doors.

Although they're not real people, smart assistants can be super helpful buddies. They're always listening (waiting for their name), and they're ready to help when you ask!

Did You Know?

Your smart assistant uses the internet to find answers and can learn new things, like you do at school!

What's one question you would ask a smart assistant?

Try it out and see what kind of fun or surprising answer you get!

Print a Toy? Yes You Can!

Imagine if your printer could make a toy car instead of printing pictures on paper. Sounds like magic, right? That's exactly what 3D printing can do!

A 3D printer takes a digital design and turns it into a real object by printing it layer by layer. Instead of using ink, it uses materials like plastic, metal, or chocolate! It slowly builds up the object from the bottom to the top, like stacking pancakes until they form a cool shape.

3D printers can create all kinds of things: toys, tools, phone cases, shoes and parts for cars and airplanes! A couple of companies are using giant 3D printers to build houses! And guess what? Chefs are experimenting with 3D printers that can print pizza and candy in fun shapes.

Doctors also use 3D printing to help people. They can print custom-made braces, hearing aids and models of organs to help plan surgeries. One day, we might be able to 3D print real body parts!

Try This: If you had your very own 3D printer, what would you create? A robot cat? A new kind of backpack? Or maybe your very own action figure?

Tech Word: 3D Printing is a way to make real objects from digital designs by printing them layer by layer.

With 3D printing, your imagination can turn into something you can actually hold!

Solar Energy Saves the Day

Have you ever felt the warm sun on your face? That sunlight is full of energy and we can use it to power all sorts of things! That's what solar energy is all about.

Solar panels are special devices that turn sunlight into electricity. You've probably seen them on rooftops or in big fields. They look like shiny black or blue rectangles, and they work best when the sun is shining bright.

When sunlight touches the solar panel, it gets turned into energy that can power lights, TVs, computers and more; all without using fuels that pollute the Earth. That means solar energy is clean and doesn't hurt the air or water.

Solar power is used in so many cool places. Some houses, schools and cars run on solar energy. There are solar-powered robots, bikes and chargers for phones. Even astronauts use solar panels to power the International Space Station while orbiting Earth!

Fun Fact: There are several calculators that use tiny solar panels to work. No batteries needed!

One of the best parts about solar energy is that it's renewable. That means we can use it again and again because the sun isn't going to run out anytime soon!

Challenge:

Design your dream solar gadget. Would it be a solar-powered backpack? A solar fan hat? Or maybe a floating solar boat?

Tech Word: Solar Panel - A special panel that collects sunlight and turns it into electricity. With solar energy, the sun is your power buddy!

Figure 6 - Solar Panels

What's Next?

Look around you.

Almost everything you see was invented by someone. From the wheel to the light bulb, from computers to robot pets, inventions have made life easier, safer, and more fun!

Long ago, people had no phones, no internet, and no video games. Can you believe that?

Then smart inventors came along and changed the world with amazing ideas. Someone dreamed of flying, and now we have airplanes. Someone else wanted to talk to people far away, and now we have smartphones and video calls!

Today, we have robots that can help doctors, cars that drive themselves, and glasses that show virtual worlds.

So, what's next?

Maybe a backpack that flies? A robot chef that makes your favorite snack? Or a homework helper bot that knows every answer?

Guess what?

The next big invention could come from YOU! Every inventor starts with a question or a problem they want to solve.

They imagine something new, sketch it out and try again and again until it works. It takes curiosity, creativity and courage; not magic!

Tech Star Challenge: Grab your pencil, crayons or markers and draw a picture of a new invention you'd like to create.

Would it help people? Would it be just for fun? What would it be made of?

Remember: The future is full of possibilities. With your imagination and a little science and tech, anything is possible. So, dream big because the world needs your ideas!

Chapter 2:
Gadgets and Gizmos

What in the World Is a Gadget?!

A gadget is a small tool or device that helps us do something cool or useful. Gadgets can be toys, tools, or tech helpers — like a smartwatch or a drone! They're often powered by electricity or batteries and designed to make life easier or more fun.

Gadgets come in all shapes and sizes. Some fit in your pocket, like a keychain flashlight, while others sit on your desk, like a 3D pen. Some gadgets make us laugh, like talking plush toys and others keep us safe, like smoke detectors. From blenders that make smoothies to electric toothbrushes that buzz your teeth clean, gadgets are part of our everyday lives.

Gadgets can be high-tech, like a drone that follows you around, or low-tech, like a handy clip-on reading light. New gadgets are invented all the time and many are made to solve problems or save time. Certain ones help people with disabilities, like voice-activated devices that can open doors or turn on lights.

Take a look around your home or classroom. You'll find gadgets everywhere – in your kitchen, backpack, or on your wrist. What gadgets do you use every day? Which one is your favorite?

Tech Word: Gadget - A clever device that does something special or fun.

Smartwatches Are Tiny Wrist Computers!

Smartwatches are tiny computers you wear on your wrist. They tell time like regular watches, but they can also do so much more! They count your steps, show your messages, track your heart rate and certain ones can let you talk to people.

Modern smartwatches come with colorful touchscreens, fun watch faces and cool apps. Want to know the weather? Tap your watch! Need a reminder to drink water or get up and stretch? Your smartwatch can buzz your wrist. Some track how well you sleep and give you tips to feel better.

There are smartwatches made for kids, with bright designs, simple games, cameras and GPS tracking so parents can find out where you are. These watches also have safety features like SOS buttons to call for help.

Many smartwatches can connect to phones, play music through wireless headphones and measure how fast you're running. Athletes use them to track workouts and students can use them as alarms or mini-planners.

Smartwatches are fun, helpful and always close by. It's like having a little assistant on your wrist that helps you stay active, safe and on time!

Figure 7 - Smartwatch

Sky Robots You Can Fly!

Drones are amazing flying gadgets that zoom through the sky like tiny robot airplanes. You control them with a remote or a smartphone. They come in all shapes and sizes.

From small sizes that fit in your hand to large ones that carry cameras and equipment.

Drones are used in lots of exciting ways. Filmmakers use them to get cool shots from the sky. Farmers use them to check crops. Firefighters use them to find people during emergencies. A few drones deliver packages or check traffic!

Many drones have special cameras that can take videos and pictures from high above. They can follow you while you run, bike, or skate, capturing awesome action shots. They use GPS to know where they are and sensors to avoid crashing into trees or buildings.

For fun, there are racing drones that zoom around obstacle courses at lightning speed. There are also toy drones with colorful lights, flips and tricks. And don't forget underwater drones that explore oceans and lakes!

Flying a drone is a fun way to learn about technology, science and how things move. You can practice steering, capturing video, or coding drones to fly on their own!

Fun Fact: Some drones are used by rescue teams to find people in hard-to-reach places!

Figure 8 - Drone

The Magic of Microwaves

Microwave ovens are magical kitchen gadgets that heat food super-fast! But how do they work? They use invisible waves called "microwaves" – a kind of energy that makes water molecules in food move quickly. This movement makes the food warm up from the inside out.

Microwaves were actually discovered by accident! A scientist named Percy Spencer realized a chocolate bar in his pocket had melted as he was working with radar. That led to one of the coolest inventions ever for the kitchen!

Today, microwaves are in homes, schools, offices and hotel rooms. They're great for heating leftovers, making popcorn, melting butter, or cooking quick meals. Certain ones have special buttons for pizza, soup, or frozen veggies.

Microwaves are fast, easy to use and don't need a stove or open flame. Just put your food in, shut the door, press a few buttons and "ding!" – it's ready.

Did You Know? Microwave-safe containers are important. You should never put metal in a microwave – it can cause sparks!

Try This: What's your favorite thing to heat up in a microwave? Popcorn? Mac and cheese? Leftover pizza?

Buttons, Joysticks and Game Power!

Gaming controllers are awesome gadgets that let you explore magical worlds, score goals and battle monsters; all with your thumbs! They're the tools gamers use to control what's happening on screen.

Controllers come in many styles and shapes. Some have buttons and joysticks. Others have touchpads or motion sensors that let you swing, shake, or move your body to play! The Nintendo Wii controller changed gaming by letting you play tennis or bowl by moving your hand.

Modern controllers often vibrate when something exciting happens, like when you crash in a race or get hit in a game. That's called "haptic feedback."

There are different controllers for each system, like PlayStation, Xbox and Nintendo Switch. Some games let you use your phone or tablet as a controller, which makes gaming more flexible and fun.

And it's not only for games! A few controllers are designed to help people with disabilities enjoy playing too. Special adaptive controllers let more kids join in the fun, no matter how they move.

Gaming controllers are more than plastic buttons. They're your gateway to adventure, creativity and teamwork. What's your favorite controller or game?

Tech Word: Controller - A device you use to tell a game what to do.

Say Cheese! A Snapshot Through History

Cameras have come a long, long way! They help us capture memories, tell stories and see the world in new ways. The first cameras, called "camera obscuras," didn't take pictures; they helped artists draw. Later, the very first real photos took hours to capture and came out black and white. People had to sit very still, or the picture would turn out blurry!

In the 1800s, cameras got better, but they were still big and heavy. Photographers had to carry tripods, glass plates and chemicals to develop the pictures. Taking a family photo was a big event and no one smiled because they had to sit still for so long!

Fast forward to the 20th century, cameras became smaller, faster and easier to use. Color photos became popular, and film cameras could be taken on vacations, to birthday parties, or underwater. Then came instant cameras like Polaroid, which gave you your picture in seconds!

Today, cameras are tiny and powerful. They're built into smartphones, laptops, tablets and toys. You can take selfies, group shots, slow-motion videos, or time-lapses in a tap. Modern cameras use digital sensors instead of film and can store thousands of photos.

Many cameras have special features like night vision, zoom lenses, or 360-degree views. There are also smart cameras that follow your face and focus automatically.

Say Cheese!

The Math Magicians in Your Pocket

Before calculators, people had to do all their math in their heads or use tools like the abacus. An abacus is a simple counting frame with beads that slide on rods. It's been used for thousands of years and is still used in several schools today!

As numbers got bigger and problems more complicated, people needed better tools. That's when inventors created mechanical calculators. These were big machines with buttons and gears that clicked and clacked as you turned them. They could add, subtract, multiply and divide but they were slow and heavy.

In the 1970s, the first electronic pocket calculators were made. They were the size of a book and very expensive! Over time, they got smaller, faster and cheaper. Today's calculators can fit in your pencil case or on your phone.

Some calculators are super smart. They can graph equations, solve square roots and help with science problems. Others are colorful and simple, perfect for learning basic math. Solar-powered calculators don't even need batteries!

Calculators make schoolwork easier, but they also help engineers, scientists and shopkeepers do their jobs. Even space scientists use advanced calculators to figure out how rockets travel through space.

Did You Know? The word "calculator" comes from a Latin word that means "small stone." Long ago, people used stones to count!

The Invisible Connector

Bluetooth might sound like something from a pirate story, but it's actually a wireless technology that connects gadgets without cords! You can use Bluetooth to link your headphones to a phone, your keyboard to a tablet, or your game controller to a console.

Bluetooth is named after a Viking king called Harald "Bluetooth" Gormsson, who helped unite parts of Denmark and Norway. Just like the king brought people together, Bluetooth brings gadgets together!

Before Bluetooth, devices had to be connected with wires. That meant tangled cords and messy desks. Now, with Bluetooth, you can walk around while listening to music, print from your phone, or send files from one gadget to another wirelessly.

Bluetooth works by using radio waves to send small amounts of data over short distances. It's perfect for personal gadgets that are close together. And it doesn't use much energy, so your batteries last longer.

There are Bluetooth speakers, earbuds, watches, fitness trackers and toothbrushes! Several toys and robots use Bluetooth to connect to apps that let you control them.

Challenge: Take a tour of your home and count how many Bluetooth devices you can find. You might be surprised!

Bluetooth makes life more convenient, less cluttered and way more fun.

Meet the Dust-Busting Machines

Vacuum cleaners are fantastic cleaning gadgets that suck up dust and dirt to help keep your home clean. But did you know the first vacuum cleaners were so big they had to be pulled by horses?

In the early 1900s, vacuum cleaners were huge machines carried on wagons. They used long hoses that went through windows to clean carpets inside homes. Only very rich people could afford them. Later, smaller, portable vacuums were invented that more people could use.

Then came electric vacuums with powerful motors and replaceable bags. These made housework faster and easier. People no longer had to sweep every corner with a broom!

Today, vacuum cleaners come in all kinds of shapes and styles. Some are upright, others are handheld, and some are built into walls. But the coolest ones are robot vacuums. These clever gadgets drive around your home on their own, dodging furniture, picking up crumbs and return to their charger when they're done.

Robot vacuums use sensors to "see" their surroundings and follow maps of your rooms. A few of them connect to your phone so you can start them while you're at school or playing outside!

Tech Word: Automation: When machines do tasks on their own without needing a person to control them.

Vacuum cleaners are a great example of how an idea as simple as cleaning can turn into a super-smart gadget.

Tiny Tech That Takes Your Temperature

Thermometers are tools that measure temperature. A long time ago, people used mercury thermometers, which had tiny tubes filled with a shiny liquid called Mercury. As the mercury got warmer, it rose up the tube to show the temperature. These thermometers worked well but were slow and could break easily.

Digital thermometers changed everything! They are faster, safer and easier to read. Instead of using Mercury, they use electronic sensors to detect temperature. The result shows up on a screen in a few seconds.

There are many types of digital thermometers. Some go under your tongue, many scan your forehead, and others go in your ear. Hospitals, schools and homes all use them because they're so convenient.

Even cooler, there are certain thermometers that are smart, which connect to phones or apps and keep track of your temperature over time. Your parents and doctors can use the information to help keep you healthy.

Digital thermometers are also used in cooking, weather stations and science experiments. They help chefs make sure food is cooked right and help scientists study the Earth's climate.

Did You Know? Smart thermometers can send your temperature to your doctor with the push of a button!

So, next time you use a digital thermometer, remember this tiny gadget is packed with high-tech magic that helps keep people safe and healthy.

Talk to Your Tech: Meet Smart Speakers

Smart speakers are like robot friends you can chat with! Just say "Hey Siri," "Alexa," or "Hey Google," and you can ask them questions, play your favorite songs, set a timer, or get a silly joke.

These gadgets have tiny microphones and speakers. When you talk, they listen and understand your words using something called voice recognition. They're connected to the internet, so they can look up answers, check the weather, or tell you what's on your calendar.

You can also use smart speakers to control other gadgets. Want to turn off the lights, change the temperature, or lock the door? Just ask! If your home has "smart" devices, the speaker can be the boss of them all.

Families use smart speakers in the kitchen to read recipes, in the living room to play music and in bedrooms for bedtime stories. Some speakers come with screens that show videos, pictures and more.

And guess what? Smart speakers learn! The more you use them, the better they get at understanding your voice, remembering your favorite songs, or giving you the news you like best.

Fun Fact: Smart speakers learn your voice and can remember your **favorite** songs.

Books, Games & Movies… All in One Gadget

E-readers and tablets are amazing gadgets that bring books, games and videos to your fingertips. An e-reader is a special kind of device for reading books. It has a screen that looks like real paper and is gentle on your eyes, even in bright sunlight.

With an e-reader, you can have hundreds of books in something as light as a notebook! You can adjust the font size, highlight words and use a built-in dictionary to learn new words as you read.

Tablets, on the other hand, do a whole lot more. You can read books, yes, but you can also play games, draw pictures, watch videos, listen to music and learn with educational apps. Several tablets come with a stylus, like a special pen, that lets you write or draw on the screen.

E-readers are great for bookworms, while tablets are super for all kinds of learning and fun. Several schools use tablets instead of textbooks to help students with interactive lessons.

Parents love tablets because they can be used for travel, learning and playtime. Just don't forget to take screen breaks to rest your eyes!

Challenge: Find one book on your tablet and read for 10 minutes!

Figure 9 - E-reader

Wristbands That Cheer You On!

Fitness trackers are smart wristbands that help you stay active and healthy. They count how many steps you take, track your heart rate and tell you how well you sleep at night.

Fitness trackers work with sensors. Tiny parts that notice movement, light, or temperature. These sensors send information to the screen or an app on your phone so you can see how much exercise you've done each day.

There are certain fitness trackers that can remind you to move if you've been sitting too long. Others let you set goals like "walk 5,000 steps" or "drink 8 glasses of water." When you reach your goals, they buzz or light up to cheer you on!

Athletes use them to track running, biking and swimming. You can use special ones designed for kids like you with fun colors, games and stickers for reaching goals. Some let you compete with friends to see who's more active.

Fitness trackers also track your sleep. They can tell if you're tossing and turning or sleeping like a rock. This helps you get better rest and feel great in the morning.

Tech Word: Sensor - A tiny part inside a gadget that detects movement, light, or heat.

Staying healthy can be fun when a fitness tracker is cheering you on!

Power Up Again… and Again!

Batteries are the secret power behind most gadgets. They give energy to your toys, remote controls, flashlights and phones. But instead of throwing away batteries when they run out, what if you could use them again and again? That's where rechargeable batteries come in!

Rechargeable batteries store energy and can be plugged in to get more. After charging, they're ready to power your gadget again. This saves money and helps the planet by creating less waste.

There are many kinds of rechargeable batteries. Some are built into devices like tablets and phones. Others look like regular batteries and can go in toys, cameras, or game controllers.

Electric cars use super-powerful rechargeable batteries to zoom down roads without using gas. Solar-powered gadgets use sunlight to recharge their batteries naturally.

Recharging is easy. You plug the battery or the device into a charger, wait a while and... zap! it's full of energy again. Many chargers have lights to show when the battery is ready.

Did You Know? Some tiny hearing aids use batteries smaller than a pea!

So don't toss the battery next time your favorite gadget runs out of juice. Recharge it!

The Coolest Gadget Ever - What's Yours?

Gadgets come in all shapes and sizes and some are just plain awesome! Maybe you've seen a toy robot that talks, a flying camera drone, or a pen that can draw in 3D. What's the coolest gadget you've ever used?

Cool gadgets solve problems in fun ways. A gadget could help you do homework faster, play music louder, or make cleaning your room easier (wouldn't that be great?).

Some gadgets help people with disabilities, such as voice-controlled wheelchairs or glasses that read aloud. Others explore space, dive deep underwater, or rescue people from danger. Technology isn't just fun. It's helpful and lifesaving!

Maybe you want to invent your own gadget. What would it do? Would it be super tiny or big like a robot? Would it be powered by solar energy, voice commands, or magic buttons?

Tech Star Challenge: Draw a new gadget you would invent. What does it do and how does it help people?

Whether it's for fun, learning, or helping others, the coolest gadgets are the ones that make life better. Maybe your invention will be the next big thing!

Chapter 3:
Robots and Artificial Intelligence

Robots Come in All Shapes

A robot is a special kind of machine that can do work. Some robots follow instructions, and some can make decisions all by themselves! What makes robots really exciting is that they come in all shapes and sizes. Some look like people with arms and legs, while others look like animals like dogs or dinosaurs. Many robots look like machines with wheels, blinking lights and moving parts, but they all have one thing in common: they are built to help!

There are tiny robots that can fit on your finger and giant robots used to build airplanes. Some robots fly through the sky, some roll on wheels and others walk or climb. Some swim underwater!

Robots can be made from plastic, metal, or soft, squishy materials. Some have sensors that help them see, hear, or feel things around them. These sensors help the robot know what to do next. For example, a robot with a camera can "see" where it's going and a robot with touch sensors can know if it bumps into something.

From robot toys that play music to giant robot arms that lift heavy cars, there's a robot for almost every job. Some are friendly and fun. Others are strong and super smart!

Tech Word: Robot; A machine that can do work, often by following instructions or using sensors to make choices.

Meet the Robots That Make Life Easier

Robots may sound like something from the future, but they're already helping us every single day. You might even have one in your home right now! Some robots can vacuum the floor all by themselves. They roll around the house, avoid bumping into furniture and clean up dust and crumbs while you play or do homework.

In big factories, robots are super helpful workers. They help build cars, toys and smartphones! These factory robots are high-speed and accurate. They can do the same task over and over without getting tired or bored. That helps humans work more safely and quickly.

Hospitals also use robots to help doctors and nurses. Some robot arms are used in surgeries. They can move very carefully and help fix parts of the body with tiny tools. There are robots that carry medicine or deliver meals to hospital rooms!

In schools, some kids learn with robot tutors. These helpful machines can talk, answer questions and also play learning games to make school fun.

And guess what?

Robots are also used in dangerous places. Some go into volcanoes or help rescue people after an earthquake. They can go where it's too risky for humans.

Robots are becoming part of our everyday world. They make work easier, faster and sometimes more fun!

Did You Know?

Some robot arms are so gentle they can crack an egg without breaking the yolk!

Robots in Space? Yep! They're Exploring Mars!

Did you know that robots help scientists explore space? Space is too far and too dangerous for people to go everywhere, so we send robots instead! These special space robots are called rovers. They travel to planets like Mars to learn more about them.

Two famous space robots are **Curiosity** and **Perseverance**. They drive around on the surface of Mars, kind of like a car with cameras and science tools. They take pictures, study rocks and look for signs of tiny life from long ago!

These space robots are very smart. They can send messages all the way back to Earth. Scientists use the information to learn more about space and how other planets work. Some rovers have robot arms that can dig in the ground and study dirt.

Since there are no people around, the rovers have to do a lot on their own. They use solar panels to get power from the Sun. At night, they rest and charge up for the next day.

Fun Fact:

The Mars rover Curiosity sings "Happy Birthday" to itself once a year! Isn't that cool?

Figure 10- Curiosity Rover

Robots Work in Factories and Farms

Robots aren't only in space. They're also busy right here on Earth. In factories, robots help build many things we use daily, like bikes, computers and candy machines! Some robots work in long lines, assembling parts super-fast and never getting tired.

Robots in factories are strong and quick. They can pick up heavy pieces, screw things together and move items from one place to another. This helps people do their jobs faster and more safely.

Robots also work on farms! Some farming robots help plant seeds, water crops and pick fruits like strawberries and apples. These robots can drive around the farm, look at the plants and decide what to do. That's a big help to farmers, especially on large farms.

There are drones, which are tiny flying robots, that fly over farms and take pictures to show which plants need more care.

Robots on farms and in factories help make sure we have the food, toys and tools we need. They do a lot behind the scenes.

Challenge: Can you spot a robot in your neighborhood? Look closely. Maybe you'll see one at work!

Want a Pet Dinosaur? Robot Pets Are Real!

Did you know that you can have your very own robot pet? That's right! Some robots are made to look and act like animals! There are robot dogs that bark and wag their tails, robot cats that purr and meow and robot dinosaurs that roar and stomp around. These robot pets are a lot of fun, and the best part is, you don't have to clean up after them.

Robot pets can move, make sounds and do tricks. Some can sit, roll over, or dance when you press a button or say a command. They can play music, light up and follow you around the room. Some even learn your name and respond when you call them.

These smart robot pets use Artificial Intelligence, or AI, to understand what you say and do. They can learn from you and get better at playing and interacting overtime. Some robot pets have sensors, so they know when you're petting them or picking them up.

Robot pets are great for kids who love animals but may not be ready for a real one. They're also helpful for people who have allergies or live in places where pets aren't allowed.

Tech Word: Artificial Intelligence (AI): A computer's way of thinking and learning like a human. AI helps robot pets learn and act smart!

Robots Have a Smart Brain!

How do robots know what to do? How can they make choices or learn new things? That's where Artificial Intelligence, or AI, comes in. AI is like a robot's brain. It helps the robot think, learn and make smart decisions - just like people do!

With AI, robots can look around, listen and understand what's happening. For example, a robot vacuum can "see" the furniture in a room and find its way around without crashing. A robot dog with AI can learn to do tricks or know when you're talking to it.

AI helps robots understand language, pictures and sounds. It teaches them to look for patterns and solve problems. The more they practice, the better they get!

Some robots use AI to learn by trying. If a robot makes a mistake, it can try again and do better the next time. Similar to when you learn to ride a bike or tie your shoes, practice helps robots too.

AI is also used in video games. The game characters you play against use AI to make smart moves. That's why they can chase you, hide, or help you during the game.

In the future, AI will help robots become smarter. They might help teach kids, talk with people, or even cook dinner.

Did You Know? Some video games use AI to make game characters act more real. That's why they seem so smart and fun to play with!

Can You Talk to Robots?

Have you ever talked to a robot? You might have without knowing it! Some robots can understand your words and talk back to you.

That's right!

They can have real conversations! These robots use Artificial Intelligence (AI) and special tools to hear, understand and reply.

Many smart robots have a microphone to hear your voice and a speaker to talk back. Some have screens that show a face or words while they speak. When you ask them a question like "What's the weather today?" they can listen, think and answer in a few seconds!

Voice robots are already in many homes. Some have names like Alexa, Siri, or Google Assistant. You can ask them to play music, tell jokes, or help you with your homework. You can ask them to set a timer or tell a bedtime story!

There are also robots made for kids. These friendly robots can help you learn math, spelling, or how to code. They can play games, sing songs and ask you fun questions.

To understand you, the robot listens to your voice and turns it into words it can read. Then, it uses AI to figure out what you mean. That's how it knows what to do next!

Some robots can talk in more than one language.

Others can tell stories with silly voices or act like your favorite movie characters.

Fun Fact:

Some robots can tell jokes, sing happy songs and also help you with tricky math problems!

Try This:

Say "Hello" to your smart speaker or voice assistant. Ask it a silly question like, "Do you like pizza?" and see what it says!

8. Can a Robot Tell if You're Happy or Sad? Yes!

Robots are getting smarter every day. Some of them can recognize emotions! That means a robot can look at your face or hear your voice and guess how you feel. Are you happy? Sad? Surprised? A smart robot might know!

Robots that recognize emotions use cameras to look at your face and microphones to hear your voice. They notice small things, like if you're smiling, frowning, or speaking in a quiet or excited tone. Then they use AI to understand what those signs might mean.

For example, if you look sad and speak in a soft voice, the robot might say, "Are you okay? I'm here to help!" If you smile and laugh, it might tell you a joke or cheer along with you!

Some emotion robots are used in schools or hospitals to help kids feel better. They might talk to children who are nervous, lonely, or scared. These robots try to be kind and friendly, like a helper or a good listener. They can be a comforting friend when someone needs one.

Other emotion robots are used to help people learn about feelings. They can show different emotions on their robot faces, such as happy, angry, or confused, so you can learn how to recognize and talk about emotions too.

Even though robots don't feel emotions like humans do, they can still help us understand our own feelings better.

Challenge:

What would you name a robot friend who cheers you up when you're sad?

What would it look like?

Draw a picture and give it a fun name!

Amazing Fact:

Some emotion robots are used to help kids with autism learn how to understand facial expressions and understand how others feel. They're like gentle, smiling teachers!

Robot Arms and Legs Are Super Strong

Robots can be super strong. Stronger than humans in many ways! Some robots have arms that can lift heavy objects, carry boxes, or move giant machine parts. Others have legs, wheels, or tracks that help them walk, roll, or climb in places people can't go.

Robot arms are often used in factories. They twist, turn and move very fast. These robot arms can pick up tiny screws or lift heavy car doors. Some are so gentle they can hold a balloon without popping it!

Robots can also have legs that walk or run. Some robot legs are used to help people who can't walk on their own. These are called bionic legs. A mix of biology (living things) and electronics (machines). Bionic arms and legs help people move, pick things up and live more freely.

Some robots have wheels to zip around quickly or track like a tank, which helps them go over rocks, mud and snow. And guess what? There are robots with wings that can fly like birds or drones.

One cool robot leg is used by firefighters. It can walk through burning buildings where it's too dangerous for people. Another robot arm is used in space, helping astronauts fix things outside the space station!

Tech Word: Bionic - A mix of biology and electronics, like robot arms or legs that help people move.

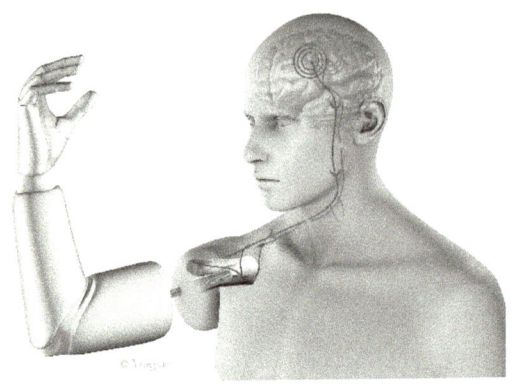

Figure 11 - Bionic arm

Think About It: If you had a robot arm, what would you use it for? Drawing? Lifting your backpack? Giving super-high fives?

Robots Go to School Like Kids Too - Kind Of!

Did you know that some robots can learn like you do? It's true! With Artificial Intelligence (AI), robots can try new things, make mistakes and get better; just like kids learning to ride a bicycle or play a new game.

These smart robots are taught by people at first. A scientist or engineer shows the robot what to do. Then, the robot practices over and over. If it makes a mistake, the robot remembers and tries a better way next time.

This kind of learning is called machine learning. It's how robots learn from experience. Like you might learn how to tie your shoes or solve a tricky puzzle, a robot learns by trying and improving.

Some robots are trained to sort toys by color, find the fastest way through a maze, or talk to people in different languages. There are robots that learn to cook simple meals or water plants. The more they practice, the wiser they get!

One super smart kind of robot learns to play games like chess. These robots think ahead, plan their moves and beat world champions!

Robots can also learn from you! Some robots made for kids can learn your name, your favorite color, or how you like to play games. Isn't that cool?

Did You Know?

Some learning robots are used to teach other robots! It's like a robot school!

Imagine This:

What if your robot could learn to pack your lunch or clean your room? What would you teach it first?

Robots Are Movie Stars

Lights, camera, robot! Did you know robots are also big stars in movies and TV shows?

That's right! Some of the coolest and funniest characters you've seen on screen are robots.

There's WALL-E, a sweet robot who cleans up trash on Earth and dreams of friendship. Then there's Baymax, the friendly, fluffy robot from Big Hero 6, who helps people feel better. And don't forget R2-D2 and C-3PO from Star Wars; they beep, talk and save the day!

Movie robots can do all kinds of amazing things. Some can fly, some are superheroes and others are super silly! They make us laugh, cry and imagine a world full of amazing robot friends.

But guess what?

A lot of these movie robots are based on real robot science! Filmmakers work with engineers and designers to make robots look and move like real machines. They study how robots walk, talk and think to make them seem real; even if they're pretending.

Some robot movie characters are played by people in costumes with special effects. Others are made using computers and animation. Either way, they help us dream big about what robots could do in the future.

Robots in movies also teach important lessons about teamwork, kindness and curiosity. Like real robots, they're always learning and trying to help.

Try This:

Draw your own movie robot character! What would it look like? What cool things would it do?

Robots Follow Instructions

Even though robots are smart, they still need help from humans. Every robot has to follow instructions to do its job. These instructions are called programs, and they tell the robot what to do, step by step.

Think of it like a recipe! Just like a recipe tells you how to make a sandwich, a program tells the robot how to move, what to say, or when to stop. If there's no program, the robot won't know what to do.

The people who write these instructions are called programmers. They use special computer languages to tell robots exactly what to do. Some programs are simple, like "move forward" or "pick up the toy." Others are long and tricky, like "Look for something blue, then dance when you find it!"

Programs also help robots stay safe. A robot in a factory, for example, is programmed to slow down if a person gets too close. A robot vacuum is told not to fall down the stairs. These safety rules are part of the robot's instructions.

Some robots can follow voice commands like "turn left" or "stop." That means people can give robots instructions by talking!

Kids can learn how to program, too! There are fun games and apps that teach kids how to give robots commands. You can make your own robot dance, draw, or play a song!

Tech Word: Program - A set of instructions that tells a robot what to do.

Challenge:

Can you write a program for a robot to brush your teeth? What steps would it need to follow?

Some Robots Are Tiny Doctors

Imagine a robot so small it can fit inside your body! Sounds like science fiction, right? But it's real! Scientists are building tiny robots that help doctors care for people in amazing ways. They are so small you can't see them without a microscope.

These tiny robots can go inside your body through a tiny tube or a pill you swallow. Once inside, they can look around, take pictures and help doctors find out what's wrong. Some can fix problems by delivering medicine exactly where it's needed!

One type of tiny robot is shaped like a little worm or fish. It swims through your body to reach places that are hard to get to with regular tools. Another type is shaped like a crab and can pinch or grab things very gently. These robots are still being tested, but they could be super helpful in the future.

Doctors use these micro-robots to do tiny surgeries without cutting open the body. That means people can get better faster and have fewer scars.

Tiny robots also help doctors check your heart, lungs, or stomach. They can take tiny samples from inside your body so scientists can study them in a lab.

Amazing Fact: Some medical robots are no bigger than a grain of rice. That's smaller than your pinky fingernail!

Imagine This:

If you had to design a tiny robot doctor, what would it look like? Would it fly, swim, or crawl?

Robots Can Build Huge Things

While some robots are super tiny, others are giant and powerful! These big robots are used to build massive things like houses, bridges, ships and skyscrapers. They use strong arms, huge tools and smart programs to help construction workers build faster and safer.

Construction robots can lay bricks, pour concrete and print buildings using 3D printing! That means a robot can squirt out layers of special building material, kind of like a giant cake decorator, to build walls one layer at a time. It's fast and very cool to watch!

Some robots are used to build homes in places where people can't easily go, like disaster zones or remote islands. They work in rain, heat, or cold and never tire. They help people stay safe by doing the hard or dangerous parts of the building.

There are also robot cranes and lifters that move heavy steel beams high up into the air. These machines are strong and steady and they follow exact instructions to put pieces in the right spot.

And guess what? Some robots are being tested to build things on the Moon or Mars one day. They could help build space stations or homes for astronauts in the future.

Challenge: Imagine a robot that could build your dream treehouse. Would it have spider legs? Jet boosters? Paintbrush arms? Draw it and explain what it does.

Cool Fact: Some 3D printing robots can build an entire small house in one day.

Future Robots Could Do Anything

The future of robots is full of exciting possibilities. Every year, robots are getting smarter, faster and more helpful.

In the future, they could do amazing things we can only dream about today.

Imagine a robot that helps you with your homework. It could explain math problems, help you write stories, or quiz you on science facts. Or how about a robot that cleans your room, folds your clothes, puts away your toys and makes your bed!

Some future robots might become ocean cleaners. They could swim through the sea, picking up trash to help sea animals and keep the water clean. Others might be space explorers flying to other planets to look for signs of life or building robot cities in space!

Future robots might help grow food, rescue people after earthquakes, or care for pets when you're away. There are ideas for robot teachers, robot artists and robot zookeepers.

Robots will also work together with humans to make life better for everyone. They won't replace people. They'll be helpers, teammates and friends.

Kids like you are already learning how to design, build and program robots. That means you could be the inventor of the next amazing robot!

Tech Star Challenge: Draw the coolest robot you can imagine. What does it look like? What special things can it do? Can it talk? Fly? Sing? Save the planet?

Remember: The future of robots depends on curious minds like yours. Keep asking questions, keep learning and one day... you might just build the next robot superstar!

Chapter 4:
The Internet and You

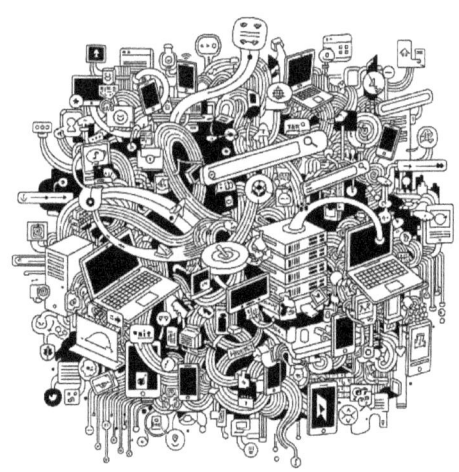

Did You Know the Internet Connects the Whole World?

The internet is like a giant invisible spider web that links computers, phones, tablets and gadgets worldwide. It helps people share ideas, videos, pictures and information at lightning speed!

Computers called **servers** help send and deliver the information you request back to your screen, and it happens so fast you barely have to wait.

The internet works day and night. Millions of people use it at the same time to play games, do homework, send photos, watch movies and talk to family and friends. It's like a magical bridge connecting everyone, from your school library to a faraway country.

The best part? You don't have to be a computer expert to use it. If you've clicked a link, watched a video, or searched for something online, you've already been a part of the internet world!

Tech Word: Internet: A global system that connects millions of computers and lets them share information.

The Internet Zooms Around So Fast

You might be wondering: how can the internet send videos, messages and games so quickly? The answer is pretty awesome! The internet uses cables underground, wires across the sky and invisible waves in the air to carry data (tiny pieces of information) from one place to another.

Any time you tap on a video or search for a game, your device sends a signal through your Wi-Fi or data connection. That signal travels to a nearby device called a **router** (kind of like a traffic controller for the internet) and then it zooms across huge networks of servers. These servers find the info you asked for and send it back to your device.

Fiber-optic cables, which are thinner than a strand of hair, carry light signals that move super-fast. Faster than cars, planes or actual sound! Satellites in space also help beam internet connections to places that don't have wires.

It's all teamwork: cables, towers, satellites, routers and computers working together to send and receive messages almost instantly.

Tech Word: Wi-Fi: This is a wireless way to connect to the internet without the use of cords.

The Difference Between the Web and the Internet

The internet and the World Wide Web are best friends, but they're not the same thing! As mentioned earlier, the internet is a giant network of computers and devices. The Web is just one part of it. It's like the internet is a big city and the Web is the library, the school and the fun park inside that city.

When you open a browser like Chrome or Safari and visit a website, you use the World Wide Web. The Web is made up of web pages filled with text, pictures, sounds and videos. These pages live on computers called web servers.

You need the internet to get to the Web, just like you need a road to get to a store. Other parts of the internet include email, apps, video calls and online games. They all travel through the internet, but not all are part of the Web.

So, next time you're watching a video online, remember you're using the Web, which is riding on the power of the internet!

Tech Word: World Wide Web is a system of online pages and websites connected through the internet.

Can Messages Really Travel in Seconds?

Yes, they can! Thanks to the internet, messages can travel across the world in just a few seconds. That means you can send a photo to a friend on the other side of the planet, and they'll get it almost instantly!

Long ago, people had to send letters by horse or boat. Later, there were mail trucks and airplanes. That could take days or weeks. But now, with e-mail and messaging apps, sending messages is super-fast.

E-mails are digital mail. You type your message, add a name or e-mail address, tap send and off it goes! Messaging apps, like WhatsApp or Telegram, do the same, but even quicker. Additionally, you can send voice notes, pictures, videos, emojis and stickers to make it more fun.

And don't worry, these messages don't get lost in the sky. They follow special paths through the internet and arrive in your friend's inbox or chat app.

Fun Fact: Over 300 billion e-mails are sent every single day!

Search Engines Know

Search engines like Bing, Google or DuckDuckGo are like super-fast librarians. They help you find the exact answer to your question from millions of websites!

For instance, when you type something into a search box like "What do pandas eat?" the search engine looks through its huge digital library of web pages and picks the best results. It finds the most useful, clear and up-to-date pages to show you.

Search engines use robots that are known as crawlers or spiders to scan the web all day and night. These robots visit websites, read the words and pictures and save what they find. That's how the search engine knows what's out there.

They also use special rules called algorithms to figure out which pages match your question best. The better the match, the higher up it shows in your results.

So, whether you want to learn about dinosaurs, find a recipe, or do school research, a search engine can help you explore the web in seconds!

Tech Word: Search Engine is a tool that helps you find answers and websites on the internet.

Why Streaming Is Popular

Streaming means watching or listening to something on the internet without downloading it first. It's like turning on a faucet, only instead of water; you get songs, shows, videos, etc. Click play and the content starts flowing to your screen right away.

Many kids use streaming to watch cartoons, learn science and listen to music. You don't need to wait for a full download. It plays almost instantly. That's what makes it so fun and fast.

Platforms like YouTube, Netflix and Spotify make streaming easy. Tap a title and it begins. Some apps remember what you like and suggest something new. You can pause it, skip ahead, rewind to hear your favorite part again.

Streaming works through special internet connections. It breaks video songs into tiny parts, sends them to your device and plays them in order. If your internet is slow, it might pause to catch up; this is called buffering.

No need for CDs and DVDs. Everything is digital. You can also stream on phones, tablets, computers or smart TVs.

Tech Word: Streaming: Watching or listening to media online without saving it first.

Can You Really Play Games with People Far Away?

Online gaming connects players all over the world. Log in and suddenly, you're racing in Tokyo, building castles in Sweden or solving puzzles with a teammate in Kenya.

Games like Minecraft, Roblox and Fortnite allow kids to team up, compete and build together no matter where they live. Some games let you chat, others let you send emojis. Others use voice chat to talk, like a phone call.

Playing online is more than just fun. It teaches teamwork, problem-solving and creativity. Players plan strategies, trade virtual items and build entire worlds. Some games have events and tournaments with hundreds of players.

To keep things fair and safe, many games have rules, privacy settings and parental controls. It's important to play responsibly; stick with friends you know and follow the game's safety tips.

Tech Word: Multiplayer is a game that lets many people play together using the internet.

How Does the Internet Help Us Learn?

Need help with homework? Curious about volcanoes? Want to hear a story read aloud? The internet is full of ways to learn.

Educational websites offer lessons, videos, quizzes and games. Kids can watch science experiments, listen to history stories or practice math skills. There are websites for every subject you can think of.

Many teachers use the internet in class. They show videos, share assignments and let students work together online.

Some schools give kids tablets or laptops to explore learning apps and digital books.

Apps like Khan Academy, National Geographic Kids and PBS Kids are made just for students. They're colorful, interactive and packed with fun facts. If you enjoy reading, online libraries have thousands of books to explore.

Learning online can happen at school, at home and anywhere else. Whether you're watching a lesson video or solving a riddle on an app, your brain gets a workout!

Tech Word: E-learning means learning through websites, apps and online classes.

Is It Really Like Talking Face to Face?

Video calls make it possible to view and speak to someone anywhere on Earth. Open an app, tap a name and suddenly, you're face-to-face, even if they're miles away.

Families use video calls to stay close. Friends use them to chat, show drawings or help with homework. Teachers use them to teach lessons from across town or across the globe.

Video calling apps like Zoom, Google Meet and FaceTime use your camera, microphone and internet connection to make everything work. You see the other person on your screen and they see you.

It feels like a regular conversation because you can wave, laugh or hold up a pet to say hi. Some apps let you add backgrounds and fun effects to your screen.

Just remember: video calls use data, so you'll need a strong internet signal. And like any online activity, it's smart to ask an adult for help and use safe apps.

Tech Word: A Video Call is known as a real-time conversation using the internet, video and audio.

Where Do All My Online Files Go?

When you save a photo or document to the internet, it doesn't just disappear into space. It goes to a special kind of computer called a server. These computers are stored in giant rooms filled with blinking lights and fans to keep them cool. This place where your files are stored is known as the cloud.

Cloud storage lets you access your stuff from almost any device. Log in on your tablet, phone or computer and there it is: your schoolwork, games or pictures. It's like having a secret locker that follows you everywhere.

One cool part is that you don't need a flash drive. If your device breaks, your files are still safe in the cloud. Many schools, businesses and video games use cloud storage to keep things running smoothly.

Remember: the cloud still lives in real computers somewhere in the world, usually guarded and protected, so your data stays private.

Tech Word: Cloud Storage involves saving your files on the internet so you can reach them from any device.

Why It Is Important to Stay Safe Online

The internet is a fun place to explore, but it's important to use it wisely. The way you'd wear a helmet when riding a bike, is the same way you need to protect yourself online, too.

Never share personal details such as your full name, address, phone number and parents' information. Keep passwords private and don't click on links from strangers. If someone online makes you feel uncomfortable, tell a trusted adult right away.

Some websites look friendly but might have tricky ads and bad information. Think before you click. Ask questions if you're unsure. It's okay to say "no" and log off.

Many sites made for kids are designed with safety in mind. They use filters, blocks and time limits. But staying safe online also means being kind. Don't say mean things and share messages that could hurt someone's feelings.

Being safe helps you enjoy the internet while avoiding trouble.

You Can Use the Internet Anywhere

Yes, and here's how! The internet doesn't just live inside computers. It travels through air, wires and space. This makes it possible to go online almost anywhere you are.

Thanks to Wi-Fi, you can use your device without plugging it in. Wi-Fi works like invisible waves that float through the air from a small box called a router. Your smart device catches these waves and connects to the internet.

If there's no Wi-Fi around, mobile data takes over. Cell towers send signals to your device using radio waves. That's how people scroll, watch, or message while moving around.

Whether you're on a beach, the internet finds a way to follow. Just make sure you have permission and a safe connection.

Tech Word: Wi-Fi is a wireless signal that connects devices to the internet.

What Is a Digital Footprint and Why Should You Care?

Every time you go online, you leave tiny marks behind, known as digital footprints. These are bits of information websites collect when you visit, click or post something.

Your footprint includes what you search, watch and share. It's not bad. It helps apps remember your settings or favorite shows. But it's important to know those actions can be seen by others, especially if your account is public.

Be thoughtful about what you post. A funny photo today might not seem funny tomorrow. Once it's online, it can be hard to delete completely.

Try this: before you click, ask yourself, "Would I want my teacher or grandma to see this?" If the answer is "maybe not," it's best to skip it.

Staying aware of your digital footprint helps you build a kind, smart online self.

Tech Word: Digital Footprint is the record you leave behind from everything you do online.

Internet Cables Run Across the Ocean Floor

Imagine super-long, super-fast wires hiding deep under the ocean! These **special wires** connect big pieces of land, like continents, just like giant spiderwebs. They zoom **phone calls, emails, and videos** to your friends and family, even if they live thousands of miles away! These wires are made of tiny glass threads, thinner than one of your hairs, but each one can carry millions of messages in just one second. Wow!

To keep these important wires safe, clever engineers wrap them in layers of strong steel, plastic, and waterproof stuff. Special boats, called **cable layers**, slowly sail across the ocean, letting the long wires sink to the bottom. Close to the beach, the wires get buried under the sand so they don't get tangled in boat anchors or fishing nets. But in the really deep parts of the ocean, they just rest on the rocky bottom, where almost nothing bothers them.

These amazing wires travel over underwater mountains, into deep ditches, and across flat ocean plains. Along the way, tiny helper boxes called **repeaters** give the messages a boost so they can zoom super far without getting tired. If a wire ever breaks (oops!), special ships find it, pull it up, fix it with a new piece, and put it back down. Fixing them can take a few days, but keeping these wires healthy means everyone around the world can keep chatting!

Today, there are over 400 of these **secret ocean wires** circling our whole planet! Together, they stretch for thousands and thousands of miles. Some connect big, busy cities, while others reach tiny islands, bringing fast internet to schools and helping doctors talk to people far away.

So, next time you watch a cartoon online or send a message to someone in another country, remember it might be diving under the ocean at super-duper speed! Those hidden wires on the ocean floor are like the **super-highways of our connected world**, carrying every "Hello!" and "Good morning!" across the big, blue depths!

The Internet Keeps Growing Every Second

Imagine a **giant, buzzing city that's always awake** – that's the internet! Every single second, new websites appear, cool videos are uploaded, messages zip around, and information zooms through invisible roads. In the time it takes you to blink your eyes, a million emails will be sent, hundreds of new pictures and posts pop up on social media, and hours and hours of new videos get added for you to watch!

This amazing growth happens because **billions of people, computers, and gadgets** join this huge worldwide network every single day. Your phone, tablet, smartwatch, and even smart sensors are always sending information. They track the weather, watch traffic, and even start your online games! Every single one of these adds to the internet, making it busier and more exciting all the time.

Think of it like this: **Search engines** are super detectives that know about trillions of web pages, so you can find answers to almost anything in a flash! And **cloud services** are like giant storage rooms in the sky, holding more information than there are grains of sand on all the beaches in the world. Every time you upload a photo, download a game, or click on something, it becomes part of this huge digital ocean that gets bigger every year! It gets so big, we even have a special word for how much it holds: a **zettabyte**, which is enough to hold every book in every library on Earth hundreds of times over!

To keep everything running, huge buildings called **data centers** are popping up all over the world. They're like giant warehouses packed with super-computers! They use clever cooling systems to stay speedy and even use clean energy like solar power to help the planet. The information then travels far and wide through those amazing underwater cables and even satellites way up in space!

Because the internet is always growing, it's easier than ever to learn things for school, stream your favorite movies, do banking online, and chat with friends no matter where they are. This amazing, ever-growing network offers endless new things to discover. And guess what? As new technologies come along, this digital universe will just keep getting bigger and bigger!

Chapter 5:
Video Games and Virtual Worlds

How Did Video Games Begin?

Have you ever wondered where video games started? Believe it or not, the very first video games weren't flashy at all. They were super simple and mostly just shapes moving on a screen! One of the earliest hits was called Pong. It came out in 1972 and all you could do was bounce a little white dot back and forth between two paddles. Think of it like playing virtual ping-pong!

Before that, computers were mostly for serious work like math and science. But inventors loved having fun, too. Early game-makers realized computers could do more than just numbers. They could also create games to play. Players loved it. Even though the graphics were blocky and the sound was just a few beeps and boops, these games felt magical.

As the years went by, new games appeared with color, better sound and fun characters like Mario and Pac-Man. Video games spread from special machines in arcades into people's homes on game consoles and computers. Pretty soon, kids everywhere had controllers in their hands.

Those first simple games showed everyone that video games could make us smile, challenge us and take us to new worlds. Without those tiny, pixel-filled games like Pong paving the way, today's super-realistic video games wouldn't exist at all!

Did You Know?

The very first video game ever was called Tennis for Two and was played on a tiny screen shaped like a green radar.

Game Consoles Are Minicomputers

Have you ever played a game on a PlayStation, Xbox, or Nintendo Switch? These machines are called game consoles and they're like minicomputers made just for games. Every console has its own tiny computer brain that understands the game and tells your screen what to show.

When you put a game disc inside or download a game, the console reads it like a storybook. Except this story is one you can control. The game might take you into a fantasy world, make you race around a track, or let you build a virtual city. The console's computer is working super-fast to draw the game world and follow your commands.

Consoles also have special chips inside to make the game look good and sound exciting. Some consoles can actually connect to the internet so you can play games with friends who live far away!

Fun Fact:

Your game console is more powerful than some computers NASA used to send rockets to the Moon! Even though they fit in your hands or under your TV, consoles can do a million things at once. Like showing detailed worlds, playing music and sound effects and tracking every jump, shot, or dance move you make.

And the most promising part? Every few years, new consoles come out with actually better features, so the adventure keeps going!

Controllers Are Your Game Hands

When you play a video game, how do you make your character jump, run, or race? That's where a controller comes in. It's like having magic hands inside the game!

A controller is a handheld device with buttons, joysticks, triggers or touchpads. Every time you press a button, the controller sends a tiny message to the game console or computer telling it what you want to do. Push up to jump, press a trigger to shoot or twist the joystick to steer your car. It's like a secret language between you and the game.

Some controllers also vibrate when something big happens in the game. For instance, when you crash a car or when a dragon roars. That makes you feel like you're really there. Other controllers, like the ones for the Nintendo Wii, can sense your motions too. Swing your arm like a tennis racket and your character swings too!

Did You Know?

There are voice-controlled games where your voice acts like a controller; say a command like "jump" or "go faster," and the game will listen!

Controllers come in all shapes and sizes because everyone plays games a little differently. Some are big and chunky with lots of buttons, some are small and fit right into your pocket. But they all do one thing; help you become part of the game world with just a touch.

Games Use Math and Art

Have you ever noticed how real video game worlds can look? There are games where grass waves in the wind, shadows stretch across mountains and characters look just like real people. None of that happens by magic. It's all thanks to math and art working together!

Game makers use math to figure out where every tiny pixel goes on the screen. They also use math for movement. If you jump up, the game calculates the curve of your jump, so it looks smooth. And when characters fight or race, math helps figure out the speed, distance and collisions.

But games need more than math. Games also need artists! Artists draw characters, paint backgrounds and design all the tiny details like clouds, flames and funny hats. Then, computer artists use 3D software to sculpt characters and worlds that look alive.

Challenge:

Next time you play a game, look closely at the world around you. See the sunlight on the grass? Someone painted that! Notice the way characters blink or move? Math is telling them how!

And the coolest part? Game makers mix math and art together every day to dream up new worlds. That's why video games look more amazing every year. They're part math puzzle, part big creative painting and 100% fun!

Games That Help You Learn

Video games can be fun and teach you fancy skills. Many games are made just to help you learn reading, math, science and problem-solving! Let's take a look at how games can help you grow smarter:

Some games teach you words by showing pictures and asking you to match them. You might tap the correct word to name a colorful animal or guess what a silly object does. This helps you practice spelling and build your vocabulary.

Math games can turn numbers into adventures. You solve puzzles to open doors, count coins to buy in-game items and use shapes to build bridges. With each puzzle you solve, you get better at adding, subtracting and thinking logically.

Science games let you explore planets, mix chemicals in a virtual lab, design your own roller coaster, etc. You learn science facts and how experiments work, all while having a blast.

Coding games teach you the basics of computer programming. You give instructions to robots, help characters follow directions or build simple programs. This is like learning a new language for computers.

Even games made mostly for fun often include skills like strategy, memory and creativity. You plan your next move, remember secret paths, design your own levels, among other things. That strengthens your brain and makes you better at solving real-life challenges!

Try This: Next time you play, notice one new skill you learn. Maybe you practice counting, reading or planning. Games that help you learn make playing double the fun! Learning while playing is super awesome!

Welcome to Virtual Worlds

Have you ever wondered what it feels like to step inside a video game? Welcome to virtual worlds, magical places you can explore, build and meet friends from anywhere! Virtual worlds are like giant digital playgrounds made of 3D graphics. You create or choose an avatar, a character that looks and acts like you want. You can run across green fields, dive under water to swim with dolphins or zoom through space on a rocket ship!

In these worlds, you can build cool things. Grab blocks like digital LEGO bricks and stack them to make castles, spaceships or roller coasters. Some virtual worlds let you plant gardens, design houses, paint on giant canvases and do several other things. Every block you place or shape you design is saved so you can visit again.

But virtual worlds aren't just for building. They're for friends too! You can join parties, chat in fun chat bubbles, dance on beach stages or team up on adventures. Talk with your buddies using text or voice and work **together** to solve puzzles or play mini-games.

Virtual worlds also host big events. You might watch concerts with pop stars, story time with an author or attend a funny fashion show to dress up your avatar. This always happens right on your screen, with people from all over the world.

Challenge: Think of a name for your dream virtual world. What would you build first? A towering treehouse, a racing track or a floating island? Write your ideas down and share them with friends!

People Make Games for a Living

Video games are made by a group of creative people. Want to see who? Here are some cool roles:

- **Game Designers** dream up the story and decide how you play. They set goals, rules and challenges.

- **Programmers** write the codes that makes the game work. They use math and logic to bring characters and puzzles to life.

- **Artists** draw characters, paint worlds and model 3D objects so everything looks amazing and colorful.

- **Animators** make characters move and act. They draw each frame so heroes can jump, dance or cheer.

- **Sound Designers** add music and effects. They make dragons roar, swords clash and play catchy tunes.

- **Producers** help teams work together and finish on schedule. They manage ideas, budgets and timelines so the game gets released.

- **Writers** craft dialogues, names and very exciting stories. They write the words that characters say and the legends you read.

- **Testers** play the game over and over to find mistakes called bugs. They help fix those bugs so the game stays smooth and fun.

- **Voice Actors** record lines to give characters their voice. They bring your favorite heroes to life with fun and exciting sounds.

Every game you love is made by these people working together. Maybe one day you could join them, too.

Did You Know? Some game creators use special shoes to walk in place and control characters in motion games!

Try This: Which role sounds most fun? Write a story or create a picture about your dream job in game making! Have fun!

What Is Augmented Reality?

Have you ever pointed your phone at the sky and seen dancing aliens on top of your house? That's Augmented Reality (AR) at work! AR mixes digital things with the real world around you. AR makes learning feel like play.

AR works by using a camera or special glasses to look at your world. Then it places digital objects in that scene. With your phone, you might see horses galloping in your backyard or goofy hats on your friends' heads. At school, AR apps can show planets floating above your desk or fossils popping out of your books!

One of the most famous AR games is Pokémon GO. Players walk around their neighborhood to find cute Pokémon hiding in parks, streets and playgrounds. The game uses GPS and your camera to place Pokémon in real places. You have to move around in real life to catch them; just like a real Pokémon trainer!

AR isn't only for games. Some apps let you try on glasses or watches before you buy them. Others help you fix cars by showing tool names on top of engine parts. AR can show safety signs on factory floors or display words in another language on street signs.

Challenge: Download an AR app with a parent's help. Look for virtual objects in your town or at home. What funny thing will you find on your kitchen table?

With AR, the real and digital worlds blend together to make everyday life way more fun and exciting!

You Can Team Up Online

Imagine logging into a world where friends appear on your screen no matter how far apart you live. That is the magic of online gaming. Thanks to the internet, millions of players join forces. You might form a squad with family members in different cities and new pals you have never met. When you work together, you become stronger as a team. Enemies flee when you plan attacks together. As you communicate through chat and voice messages, you learn to share ideas and listen carefully.

Online games provide all kinds of adventures. You might explore enchanted forests searching for hidden treasures against mythical beasts. In a racing challenge, you accelerate down twisting tracks while cheering teammates into first place. In building games, you gather materials and construct amazing structures layer by layer. Every mission demands coordination that teaches cooperation, leadership and quick thinking.

Playing online also means meeting players worldwide. You can join friendly clubs based on hobbies such as painting, cooking and exploring virtual islands. Together, you host parties, solve puzzles and stage epic battles in massive arenas. Team spirit grows when each member contributes special skills toward a common goal. As you celebrate victories, you also support one another in challenges.

Challenge: Invite three friends to play an online game this week. Coordinate a plan, assign each friend a task and complete a mission together. Notice how teamwork makes every adventure more fun and exciting.

Grown-Ups Play Too

Video games welcome players of every age. Parents, teens and grandparents often pick up controllers to join the fun at home and online. After a busy day, many adults relax in virtual worlds. Alongside friends worldwide, they explore new quests and challenges. They laugh with friends while solving puzzles and racing at top speed.

Some grown-ups go into strategy titles. They build kingdoms, manage resources and lead armies. Many adult gamers also play cooperative missions that require teamwork and creativity. Others prefer puzzle adventures where clever thinking unlocks secrets. Racing fans zoom through tracks chasing new records. Creative players shape colorful buildings that rival LEGO masterpieces.

Playing together brings families closer. When you teach your dad simple moves, you both practice patience. If Grandma joins a mission, her determination shines as she overcomes challenges. Shared victories create memories to treasure. Laughter fills the room when a surprise win catches everyone off guard.

Video games also spark stories from the past. Parents recall arcade games with pixel heroes. You compare their old adventures with today's lifelike characters. That tale-sharing makes family time extra special and helps you see how games have grown.

Fun Fact: Globally, the average gamer is about thirty-five years old today. Many adults love gaming just like kids!

Try This: Pick a game you enjoy and teach an adult a special trick. Then switch roles so they show you a move. Celebrate the teamwork and smiles you create together.

You Can Make Your Own Game! Here's How

Have you ever dreamed of building your own video game? Simple tools can help you start today. Scratch, along with Roblox Studio lets you design mini-games with block-based commands and user-friendly scripts. Both provide step-by-step tutorials that teach movement, actions and interaction in fun lessons.

First, grab a pencil and notebook. Sketch your game's main character, the world it lives in and the goal players must reach. Maybe players collect stars in outer space, guide a robot through a maze and solve mysteries in a haunted mansion. Writing your ideas gives you a clear plan.

Next, open Scratch along with Roblox Studio and follow a beginner lesson. In Scratch, snap colorful blocks to make sprites walk, talk and bounce. In Roblox Studio, place digital bricks and add simple scripts that bring worlds to life. Test each new move so you catch bugs early and keep the fun flowing.

When your game feels ready, it's time to share. Click Share on Scratch then Publish in Roblox Studio so friends can play. Ask players to try levels and send suggestions. You can then polish graphics, add fresh features and fix issues based on feedback. This process helps you become a better creator.

Be proud of your first game! Every professional developer began with small projects. With curiosity, imagination and practice, you can turn simple ideas into awesome adventures that others will love.

Share your game worldwide.

You Control Your Avatar

Stepping into a game means taking charge of an avatar, a character you guide through every adventure. You pick its skin tone, hairstyle, costume and accessories to match your style. As you complete quests, find hidden items and earn badges, you unlock new outfits that show off your achievements. Friends spot your team at a glance when you wear matching colors.

Your avatar expresses feelings through emotes; short actions let you wave, cheer, dance and laugh. Send a happy dance after a great win or show surprise when you uncover a secret chamber. Those moves add personality without a single line of text.

Playing also makes your character stronger. Each challenge you overcome grants experience points and new skills, such as higher jumps, faster swims and powerful strikes. Leveling up highlights your progress and sparks excitement for each next mission.

Some games let you record voice lines so your avatar speaks with your own tone. That brings your hero to life and makes every moment unique.

Try This: On paper, draw an avatar for a jungle explorer, deep-sea diver or space pilot. Sketch the outfit, list three special moves and imagine one hidden treasure it must find. Share your design with friends and see their creations unfold.

Why Game Ratings Matter

Before playing a new title, look for its rating, a symbol that guides families to age-appropriate content. Common labels include **E** for Everyone, **E10+** for ages ten and up, **T** for Teen and **M** for Mature audiences. These tell you whether a game features gentle humor, cartoon action, intense battles or realistic themes.

Ratings boards consist of parents, teachers and child-care experts who review each scene. They note adventure levels, spooky moments and any strong language. Then, they assign clear symbols to boxes and digital stores, so families know what to expect.

Using ratings makes playtime safe and fun. A higher age tag often means more complex puzzles, deeper stories or stronger action that suit older players. Guardians can approve games with confidence and help kids find new favorites at the right time.

Many regions add brief content notes such as "Fantasy Violence" or "Mild Language" to explain why a game earned its label. That extra detail helps families make smart choices.

Try This: When browsing games next time, spot the rating symbol and note its description. Ask an adult what it means before you click "Play." Good choices set the stage for the best adventures.

Games Are Fun on Purpose

Games feel exciting because creators design every part to spark curiosity, test skills and reward effort. First, they set clear missions like collecting glowing gems, rescuing friends or building thriving cities. Those missions guide players step by step.

Challenges grow tougher as you learn new moves. Early puzzles teach basics. Later tests demand clever thinking and quick action. Overcoming each challenge feels like winning a reward minted just for you.

Rewards come in many forms: stars, coins, badges and new outfits. Each prize unlocks fresh levels, special tools and hidden secrets. Seeing rewards appear keeps you eager for the next chapter.

A gripping story ties everything together. You follow heroes on epic quests, meet quirky sidekicks and uncover worlds hidden behind locked doors. A strong narrative makes every victory feel meaningful.

Designers balance missions, challenges, rewards and stories with care. Goals show you what to do. Puzzles test your skills. Prizes celebrate progress. Tales draw you deeper into the game world. That perfect blend keeps you returning for new thrills and discoveries.

Try This: Think of your favorite game. Identify its mission, challenge and reward that you enjoy most. Noticing these pieces reveals how games are built to delight players every time.

Chapter 6:
Space Tech: From Earth to the Stars

Rockets Lift Us Into Space

Rockets are super-powerful machines that push spacecraft beyond Earth's sky. Special fuel burns in engines to create a blast of hot gas inside a rocket. This blast pushes the rocket upward at incredible speed, fast enough to escape Earth's pull!

Early rockets used simple designs with a tall body, fuel tanks and an engine at the bottom. Today's rockets come in stages. The first stage lights up, carries the rocket high, then drops away once its fuel runs out. The next stage fires up, taking satellites or astronauts even farther.

When you watch a launch, you see a glowing flame and hear a thunder-like roar. The ground shakes while the rocket climbs through fluffy clouds. After a few minutes, the rocket reaches space, a place without air and gravity so low you float!

Rockets help us reach the Moon, send probes to Mars and launch satellites that beam TV shows, phone calls and weather reports back down. Space companies around the world build rockets with new tech to make launches safer, cheaper and reusable. Some rockets land on special pads so engineers can prepare them for another flight.

Fun Fact: The fastest human-made object is a rocket probe called Parker Solar Probe. It zooms so close to the Sun that its top speed hits 700,000 km/h!

Next time you see a blast-off on TV, picture the teamwork behind each launch: from scientists who design engines to astronauts who ride inside. Rockets turn dreams of space travel into thrilling reality!

Satellites Fly Above Our Heads

Satellites are like floating helpers circling high above Earth. They ride rockets into space, then switch on solar panels to soak up sunlight for power. Once in orbit, a satellite travels around Earth at thousands of kilometers per hour. From that height, it sees half the planet at once!

Weather satellites watch clouds gather and storms form. They send pictures to meteorologists who forecast rain, wind and snow. Earth-observation satellites snap detailed photos of forests, farms and coastlines. Scientists use these images to track changes in nature, from melting ice to growing cities.

Communication satellites beam phone calls, TV shows and internet data across oceans and mountains. When someone talks on a satellite phone, their voice flies up to space, bounces off the satellite and returns to a receiver halfway around the world. That's how distant friends stay in touch!

Some satellites keep an eye on space itself. They spot incoming asteroids, measure cosmic rays and study the Sun's activity. Space weather satellites warn us about solar storms that could disrupt power grids on Earth.

Each satellite has special instruments: cameras, antennas or sensors that detect heat, light and magnetic fields. Engineers design satellites to withstand harsh conditions; extreme temperatures, vacuum and tiny particles flying at high speed.

Try This: Draw your own satellite. Give it solar panels, a camera and antennas. Then explain what your satellite will study or help with up in space!

Space Stations Are Homes in Orbit

Imagine living in a house floating high above Earth. That's what a space station is. A laboratory and home where astronauts live and work for months at a time. The biggest one today is the International Space Station (ISS). It circles Earth every 90 minutes at about 400 km up, so residents see sixteen sunrises each day!

Inside the station, air is just like on Earth. Special filters clean it and keep oxygen levels safe. Astronauts move around with Velcro patches on their socks and magnets on their tools, so nothing floats away. They sleep in cozy sleeping bags attached to the wall.

Space stations carry science labs loaded with experiments. Astronauts grow plants to learn how green beans sprout without heavy gravity. They test new materials that could help build better spacecraft. They even study how the human body adapts to low gravity useful for future missions to the Moon and Mars.

Living in orbit teaches teamwork. Engineers on Earth guide astronauts by video chat, helping them fix broken parts or run tricky tests. International crews from different countries share meals, games and stories, proving that people can cooperate even when they float far from home.

Resupply ships arrive every few months with fresh food, water and new experiments. When a spacecraft docks, astronauts open the hatch and unload crates full of supplies. Then they send old gear back to Earth for repair.

Did You Know? The ISS is so big that astronauts aboard can see houses, roads and even large ships moving on Earth below!

Telescopes See Farther Than Ever

Telescopes are giant eyes pointed at the sky. They collect light from stars, galaxies and nebulae to help us understand the universe. Ground telescopes sit atop mountains to escape city lights. Space telescopes orbit above Earth's atmosphere for crystal-clear views.

A simple telescope uses a curved mirror or lens to gather light. Bigger mirrors collect more light, revealing faint objects billions of light-years away. Space telescopes like Hubble capture stunning images of glowing clouds where new stars are born and ancient galaxies that formed long ago.

Modern telescopes use cameras that detect colors beyond human vision, which are infrared heat rays and ultraviolet light. Infrared telescopes spot cool objects such as newborn planets still surrounded by dusty disks. Ultraviolet telescopes study hot stars and violent explosions called supernovae.

Some telescopes pair up across continents to act as a single giant eye. Radio telescopes listen for signals from black holes and distant galaxies. By combining data from many dishes, astronomers create detailed radio pictures of space.

Telescopes also help find new worlds. The Kepler space telescope measured tiny dips in starlight when planets passed in front of their stars. That revealed thousands of exoplanets, planets beyond our solar system.

Each telescope requires careful design: mirrors polished to perfection; detectors cooled to near absolute zero and computers to process massive data streams. Teams of scientists analyze those data to answer big questions: How did our universe start? Are we alone?

Challenge: Grab a pair of binoculars and spot the Moon's craters. Imagine the telescope that captured the first clear photos of Mars!

Rovers Explore Other Planets

Rovers are robotic explorers that drive across alien landscapes. Engineers build these special vehicles to roam where people cannot safely travel. They land on planets with a soft touchdown then wake up their wheels. Each rover carries cameras, drills and scientific instruments to study rocks, soil and the air.

When a rover rolls onto dusty plains, its cameras take detailed pictures. Scientists on Earth receive those pictures and plan the next move. Rovers use robotic arms to scoop up soil samples and drill tiny holes in rocks. Instruments inside test for water traces and search for signs of ancient life.

Solar panels or nuclear batteries power rovers for months and years. They brave harsh conditions: freezing nights, dusty skies and sharp rocks. Wheels help rovers climb slopes and cross small craters. If a wheel gets stuck, engineers send new commands to wiggle it free.

Famous rovers include Sojourner, Spirit, Opportunity and Curiosity. Opportunity exceeded its planned 90-day mission by nearly fifteen years. Curiosity still roams Mars with its six wheels and laser that zaps rocks to reveal their secrets. The newest, Perseverance, searches for ancient microbial life and collects samples for return to Earth.

Every new rover teaches us about other worlds and how to build better explorers. They help plan future missions with astronauts by mapping safe routes and finding places with water ice. Thanks to these tireless robots, our knowledge of planets beyond Earth grows every day.

Try This: Draw your own rover. Give it sturdy wheels, solar panels and a robotic arm. Then name your rover and describe its first mission on a distant planet!

Astronauts Use High-Tech Suits

Astronauts wear special suits that work like miniature spacecraft. These suits keep them alive in the vacuum of space where no air exists. A suit has many layers to protect against temperature extremes, tiny particles and harmful radiation from the Sun.

The outer layer resists scratches from space dust. Beneath that, insulating layers trap warm air so the astronaut stays at a safe temperature. Seals at the wrists, ankles and neck keep the suit airtight. A helmet with a clear visor provides a wide view and shields astronauts from bright sunlight.

A life support backpack pumps oxygen into the suit and removes carbon dioxide. It also controls temperature with cooling water tubes that wrap around the astronaut's body. Special gloves let astronauts grip tools and handle delicate samples on the Moon or in space stations.

Communication gear built into the helmet allows astronauts to talk with each other and with mission control on Earth. Microphones and earphones sit inside so voices sound clear despite the noisy fans. Lights and cameras on the helmet help astronauts see in shadowed areas and record their work for scientists below.

Suits include layers that reduce sharp jolts when moving heavy equipment. Flexible joints at shoulders, elbows and knees let astronauts bend and climb ladders. Suits for the Moon missions had bright white covers that reflected heat from the Sun.

Every new mission brings suit improvements. Engineers test fabrics, seals and life support systems in large vacuum chambers on Earth. That ensures each suit will protect explorers during spacewalks, Moonwalks and future trips to Mars.

Moon Missions Changed Everything

When humans first set foot on the Moon on July 20, 1969, history changed forever. Neil Armstrong and his partner, Buzz Aldrin of Apollo 11 climbed down from their lunar module and left the first footprints in lunar dust. "That's one small step for man, one giant leap for mankind," Armstrong said as he stepped onto the surface.

Moon missions gathered rocks that scientists still study today. Those rocks revealed clues about how the Earth and Moon formed billions of years ago. Instruments left behind measured moonquakes and sunlight reflections that helped map the entire lunar surface.

The technology developed for Moon missions led to everyday inventions. Memory foam, which we use in mattresses and helmets, came from materials created to cushion astronauts. Scratch-resistant lenses for eyeglasses grew out of visor coatings. Portable cordless tools, satellite navigation systems and advanced water filters all trace back to Moon-landing research.

The Apollo program also showed people working together across nations can achieve incredible goals. Millions watched landings on television and felt proud about humans exploring a world beyond Earth. Space agencies around the globe started planning new spacecraft and missions to other planets.

Learning to live and work on the Moon taught engineers how to build habitats, life support systems and rovers.

Those lessons guide plans for bases on the Moon and Mars. Future explorers will visit again, using lunar water ice to make rocket fuel and oxygen.

Moon missions opened our eyes to what is possible when curiosity, teamwork and bold ideas come together. They sparked a love of science that continues to inspire students and inventors everywhere.

Mars Is the Next Big Goal

Mars has captured human imagination for centuries. Nicknamed the Red Planet, it shows rusty deserts, giant volcanoes and valleys carved by ancient rivers. Scientists believe Mars once held liquid water and possibly primitive life. Today Mars stands as the next frontier for exploration.

Robotic missions over the past decades have mapped its surface and analyzed its soil. Rovers have driven over dusty plains, climbed slopes and sent stunning panoramas back to Earth. Orbiters circle the planet, studying its climate and searching for water ice at the poles.

Future missions aim to land astronauts on Mars. They will live in habitats that shield them from extreme cold and radiation. Engineers plan systems to produce water, oxygen and rocket fuel using local materials. Growing food in greenhouses will let crews eat fresh vegetables on long missions.

Getting to Mars takes around six to nine months. Spacecraft must carry enough supplies and reliable life support for that journey. Engineers design heat shields to protect landers during entry through Mars' thin atmosphere. Parachutes and retrorockets will slow the descent for a safe touchdown.

Astronauts on Mars will explore canyons, climb volcanoes and search for signs of past life. Their discoveries could answer big questions about how life began on Earth. Lessons learned on Mars missions will actually help plan bolder expeditions to worlds like Europa and Titan.

Challenge: Imagine you are the first person to step on Mars. Write a short diary entry about your first day exploring the red landscape and what you hope to discover!

Satellites Help Us on Earth

Have you ever checked the weather before heading outside? That forecast often comes from satellites circling high above us. These space helpers watch Earth as if they are giant, silent guards. They spot storms brewing over oceans, track wildfires in forests and measure how much ice covers the poles. With that information, meteorologists give us accurate forecasts so we know when to grab rain boots or sunscreen.

Satellites also keep an eye on our planet's health. They count trees in rainforests, watch crops grow in farmlands and measure pollution in the air. Scientists use these data to protect wildlife, help farmers plan harvests and study climate changes over many years. Satellites warn communities about future floods and teach us how to care for Earth after spotting sea-level rise or melting glaciers.

Every time you make a phone call across continents or watch live video from halfway around the globe, you're using communication satellites. Your voice travels up in a radio wave, bounces off a satellite and comes back down to its destination. Internet, TV shows and GPS signals rely on these orbiting messengers.

Some satellites try to detect sounds we cannot hear. They pick up radio waves from lightning storms, volcanic eruptions, and even whales singing in the deep ocean. By collecting one big puzzle of signals from space, satellites help scientists solve mysteries about our planet.

Try This: Draw a picture of Earth from space with several satellites hovering around it. Label each one with its job; weather watcher, farmer's helper, or phone-call carrier. Seeing all those helpers at work makes you appreciate the high-flying friends above!

Robots Help Build in Space

When rockets reach orbit, they sometimes carry special robotic arms and free-flying robots that help build things in space. These robots work where it's risky for humans to float around. Think of them as helpful buddies who never get tired.

One famous example is Canada's Canadarm2, a giant mechanical arm on the International Space Station. It reaches out, grabs arriving cargo ships and gently snuggles them to the station's docking ports. Astronauts control it from inside, guiding it like a crane operator on Earth. Canadarm2 also helps move big solar panels into place and carries astronauts safely during repairs.

Tiny floating robots called SPHERES zip around the station's interior. They look like soccer balls with little thrusters. Engineers on Earth send them tasks, such as taking measurements or testing new sensors. Watching SPHERES fly in zero gravity is like seeing a tiny spaceship ballet in slow motion.

Future missions to the Moon and Mars will actually rely more on robots. Before astronauts arrive, rovers and autonomous machines will build landing pads, habitats and solar arrays. They'll weld metal parts, assemble tubes for growing plants and lay down paths to keep vehicles safe.

Working side by side with robots, astronauts can focus on experiments and exploration. Robots handle heavy lifting, risky repairs and routine checks. Together, humans and machines make sure space stations and future colonies stay strong and ready for adventures beyond Earth.

Fun Fact: Some space robots use extendable "tendril" fingers to grab onto handrails and tunnels, just like a gecko's feet sticking to a wall.

Space Food Is Special Science

Eating in space is more than just lunch and snacks. It's a science experiment in every bite. Without gravity, crumbs float away and liquids form floating blobs, so astronauts eat special meals that stay in place and taste great.

Early space food came in tubes and packets. Astronauts squeezed pureed beef stew straight into their mouths. It worked, but it didn't feel very cozy at dinner time. Today, space chefs pack food in sealed trays. You heat your meal, pull off the lid and use a fork with magnets so it sticks to your tray. No more floating peas!

Nutrition is key. Food must last months without spoiling, yet keep astronauts healthy and strong. Dietitians choose ingredients rich in vitamins, protein and fiber. Meals range from pasta with tomato sauce to tortillas with peanut butter. Fresh produce flies up on resupply missions, so they get crunchy apples and leafy greens every few months.

Drinks come in special pouches with straws that seal tight. There's coffee, fruit juices and lemonade. Sipping through a straw keeps the liquid from forming droplets that drift into equipment.

Eating together in the space station's "galley" is a special moment. Astronauts share stories and laugh about floating utensils and toast with juice pouches.

Try This: Create your own space menu at home. Pick five favorite foods. Think about how you would package them to keep each bite from drifting away. Share your menu with friends and see who designs the coolest space picnic.

GPS Works Thanks to Space Tech

Have you ever used a map app to find your way to a friend's house? That magic comes from GPS. It is a system of satellites that tells your phone exactly where you are on Earth. GPS stands for Global Positioning System, and it works because space technology keeps those satellites circling overhead.

At least 24 GPS satellites fly in precise orbits about 12,550 miles above us. Each one carries an ultra-accurate clock powered by atomic vibrations. They send time signals down to your device. Your phone figures out your exact latitude, longitude and altitude by calculating how long those signals take to arrive from four different satellites.

This space trick is remarkably precise, down to just a few meters! With that data, your map app can guide you around town, show traffic congestion in real time and help emergency crews find people in need. Hikers use GPS to navigate trails, pilots to chart courses and ships to sail safely across oceans.

GPS satellites also carry sensors that monitor Earth's environment. They measure signals bounced off the ground to detect changes in sea level, ice thickness and soil moisture. Farmers use this data to plan planting, and scientists track weather patterns and natural disasters.

Fun Fact: Some high-precision GPS receivers can locate you within a few inches just a little wider than a ruler!

Space Telescopes Discover New Worlds

Space telescopes are like powerful detective tools floating above Earth's atmosphere. They catch light from distant stars and planets without the blurring effects of air. With these telescopes, astronomers have found thousands of planets beyond our solar system, called exoplanets. Some are rocky like Earth, others are giant gas worlds bigger than Jupiter.

A space telescope spots exoplanets by watching a star's brightness. When a planet crosses in front of its star, the starlight dips slightly. That tiny dip tells scientists an unseen planet is passing by. By measuring how long the dip lasts and how deep it goes, they learn the planet's size and how close it orbits its star.

Other space telescopes use infrared sensors to see heat. Warm objects give off infrared light, so telescopes can detect young planets still glowing from their formation. Infrared also reveals dusty disks around newborn stars where planets form. These views help scientists understand how solar systems take shape.

Once a new world is found, researchers aim to study its atmosphere. If a telescope sees signatures of water vapor, carbon dioxide or methane, that hints at conditions where life might exist. Future space telescopes will use special mirrors and cameras to capture faint colors of exoplanet atmospheres, searching for signs of habitability.

Every new discovery expands our view of the universe. Space telescopes show that planets are common and diverse. Some worlds may have rings, some may be covered in lava lakes, others might float in eternal twilight. Each exoplanet invites us to imagine what life could be like far beyond Earth.

Try This: Pick an exoplanet discovered by space telescopes and draw how you imagine its sky, land and possible creatures.

Spacecraft Go Beyond Our Solar System

Most spacecraft stay inside our solar system, but a few have broken free and journeyed into interstellar space. The most famous are Voyager 1 and Voyager 2. They launched in 1977 with cameras, instruments and special records carrying sounds and pictures of Earth. These probes flew past Jupiter, Saturn and their moons, sending back breathtaking images of rings, storms and icy surfaces.

After completing those missions, each Voyager climbed the Sun's magnetic bubble, called the heliosphere and entered the space between stars. In this lonely region, they still send faint radio signals back to Earth. Scientists study those signals to learn about cosmic rays, magnetic fields and the thin gas that drifts between stars.

Beyond Voyager, New Horizons flew by Pluto in 2015 and it revealed mountains of ice, frozen plains and giant heart-shaped formations. It continues onward toward the outer edge of the Kuiper Belt, a territory of icy bodies left over from our solar system's formation.

Building spacecraft for interstellar travel requires careful design. They need power sources that last decades, such as plutonium-238 generators. They must handle extreme cold, vast distances and tiny particles traveling at high speed. Engineers program these probes to operate automatically, since radio signals take hours to cross the expanding gap.

Interstellar spacecraft carry messages from Earth, our first ambassadors to distant space. They remind us of humanity's curiosity and drive to explore the unknown. One day, new missions may ride laser beams or use light sails to reach nearby stars in mere decades. Until then, our farthest travelers sail on and teach us about the vast universe beyond our solar home.

Challenge: Imagine you design a spacecraft for another star. What power source would you choose? How would you send messages back home? Sketch your ideas on paper.

Kids Like You Might Go to Space Someday

Going to space used to be reserved for a few astronauts, but that is changing fast. Companies now build rockets that carry private passengers above Earth's atmosphere. Soon, families, scientists and even school groups could enjoy a zero-gravity trip.

Training begins on Earth. Future space travelers learn to handle weightlessness in special planes that fly gentle arcs, letting people float for brief moments. Schools and museums offer astronaut camps that teach rocketry, navigation and basic science. Students design experiments to fly on suborbital flights, then analyze data when they return.

In the near future, small space stations may host visitors for days at a time. You could float through corridors, conduct simple science tests and watch Earth turn beneath you. Imagine waking up to a sunrise every 90 minutes as you circle the planet.

Beyond low orbit, tourist missions will explore the Moon's surface. Companies plan lunar landers that carry crews to touch dusty plains, explore craters and collect rock samples. Training will include rover driving and habitat living skills for multi-day stays.

As technology advances, new fields open up. Space agriculture programs teach you to grow plants with LED lights and recycled water. Space manufacturing uses microgravity to create materials with unique properties. Hospitals may send medical tests to space labs for faster results.

To prepare, you can study math, physics and computer science. Clubs that build model rockets give you hands-on experience. Learning engineering skills helps you design tools that work in space's harsh environment.

One day, you may strap into a seat, feel your rocket engines ignite and watch Earth fade below. The sky will stretch out in black velvet, studded with stars. The next generation of space explorers could start right here, with dreams bigger than our home planet.

Try This: Write a letter to your future self as an astronaut. Describe what excites you most about living in space and what experiments you hope to run.

Chapter 7: Green Technology

Solar Panels Turn Sunshine into Power

Imagine a shiny glass rectangle soaking up sunlight and turning it into electricity -that's a solar panel! Solar panels are made of extraordinary materials called **photovoltaic** cells. When sunshine hits these cells, tiny particles inside wake up and start moving. This movement produces an electric current that flows via wires into your home's outlets.

On sunny days, rooftops covered in panels collect sunlight all morning until evening. Inside your house, an inverter changes the panel's direct current into alternating current, which powers lights, TVs and chargers. Any extra electricity can flow back to the power grid for neighbors to use. At night, your home draws from the grid until the sun returns the next day.

Solar panels need little care. A quick rinse with a hose takes out dust and leaves every few months. Engineers design panels to last twenty-five years or more, longer than many gadgets in your house! In sunny regions, families save money on electricity bills while helping the planet by using clean energy instead of fuels that pollute.

Did you know spacecraft, satellites and even streetlights use solar panels? Off Earth, solar power keeps instruments running for years with no gas tank in sight. Back on the ground, you might spot solar-powered backpacks that charge phones during a hike or garden lights that glow after dusk.

Try This: Draw a house with solar panels on its roof. Then sketch a day-to-night timeline showing how panels collect sunlight, power your home and refill at sunrise. Seeing each step reminds you how sunshine can brighten our lives in many ways than one!

Wind Turbines Spin Up Clean Energy

On windy hills and open plains, tall white towers hold giant blades that slice through the air. These are wind turbines. When wind pushes against the blades, they spin around a hub. That spinning turns a shaft inside the tower, which drives a generator to produce electricity.

Each blade may stretch as long as a football field. High above the ground, wind blows faster and steadier, so turbines generate more power than smaller machines. A single modern turbine can make enough electricity for hundreds of homes every year.

Inside the turbine's "nacelle," you'll find gears that increase the blade's slow spin into a fast rotation for the generator. Transformers at the base boost the voltage so power flows efficiently through transmission lines to towns and cities. Smart controls swivel the entire tower to face the strongest breeze all day long.

Wind farms group many turbines across hillsides or offshore on the ocean. Offshore turbines catch steady sea breezes and create large amounts of clean energy without occupying farmland. Because no combustion is involved, wind power produces no greenhouse gases as it spins.

Fun Fact: The oldest wind turbines appeared in Persia long ago, grinding grain with horizontal sails. Today's high-tech turbines trace their roots to those ancient windmills that harnessed the breeze.

Try This: On a windy day, make a simple pinwheel with paper and a pencil. Watch how the breeze makes it spin. Then imagine dozens of these working together to light up your school!

Electric Cars Don't Need Gas

Electric cars look like regular vehicles, but inside, they carry a big battery instead of a fuel tank. Plug them into a charging station or home outlet, and the battery soaks up electricity just like your phone. Once charged, the car's electric motor uses that power to spin wheels silently and smoothly.

Without a gasoline engine, electric cars avoid tailpipe emissions that add pollution to the air we breathe. That means zero exhaust gases leaving the car. No smoke, no smells. Owners enjoy instant torque, so electric cars zip from zero to speed faster than many gas cars. Regenerative braking diverts the motor into a generator when you slow down, sending extra energy back into the battery.

Charging stations appear at shopping centers, parking lots and along highways. Fast chargers can refill a car's battery to 80 percent in about 30 minutes. Overnight charging at home tops up the battery for a full day's driving.

Electric vehicles need less maintenance too. Fewer moving parts mean no oil changes, no spark plugs and fewer brake pad replacements. With digital dashboards, drivers can see real-time energy use and plan trips around available chargers.

Fun Fact: The first practical electric cars appeared over 100 years ago. They competed with gas cars until gasoline became cheap. Today, big battery advances mean electric cars are back, cleaner than ever.

Try This: Sketch your dream electric car. Add solar panels on the roof or wheels that make energy when rolling. Then label how it charges and keeps our air clean.

Smart Thermostats Save Energy

A smart thermostat learns how you heat and cool your home to save energy without making you feel too hot or too cold. Installed on the wall like a regular thermostat, it connects to your Wi-Fi and uses sensors, schedules and weather data to adjust temperatures automatically.

During winter mornings, a smart thermostat can raise heat just before you wake up and lower it after you leave for school. It learns your routine over a few days, then follows that pattern each week. If you forget to adjust it before a weekend trip, the thermostat switches to energy-saving mode until sensors detect someone is back.

In summer, it can lower AC use when the sun blazes through windows and raises it at night for a comfortable sleep. Many models let you control settings from a smartphone app. If a snow day keeps you home, you can tweak the temperature from bed with a tap on your phone.

Smart thermostats send reports showing how much energy you've used each day. That helps families see when they might save extra power. By cutting heating and cooling waste, households often trim energy bills by up to 20 percent. Less energy use also means fewer greenhouse gases from power plants.

Fun Fact: Some smart thermostats work with voice assistants so you can say, "Set the temperature to 70 °F," and it happens instantly.

Try This: Track your home's temperature schedule for one week. Note when heating or cooling turns on and off. Then, imagine how a smart thermostat could learn those times to save energy while keeping everyone comfy.

LED Lights Use Less Electricity

Have you seen tiny light bulbs that glow brilliantly yet stay cool to the touch? Those are LED lights, and they shine using far less power than old-fashioned bulbs. Inside each LED, special materials create light when a tiny electric current passes through. This process stays efficient. More light, less heat.

Because LEDs use so little energy, they last a very long time. A single LED bulb can glow for up to 25,000 hours. That means if you left one on ten hours each day, it would keep shining for nearly seven years! By swapping out old bulbs for LEDs at home, families cut electricity use and save money on their bills.

LEDs come in every shape and color you can imagine. You might hang glowing strands of LEDs during holidays or add a bright desk lamp that helps with homework. Some LEDs fit into garden lights powered by solar panels, so they shine at night without using a drop of household electricity.

These lights also help the planet. Power plants burn less fuel to keep LEDs lit, which means fewer pollutants in the air. Cities have switched streetlights to LED technology, making roads safer while cutting energy waste. Even huge sports arenas use LED floodlights to brighten every seat with less power.

Try This: With an adult's help, swap one old bulb in your home for an LED. Track your energy use on that light for a week. Notice how it stays cool and shines brightly while using less electricity.

Recycling Turns Trash into New Things

Imagine tossing a plastic bottle into a special bin and watching it transform into a playground swing. That magic happens through recycling. When you separate paper, plastic, glass and metal, recycling centers sort those items, clean them, then melt or mash them into raw materials. Those materials reappear as new jars, notebooks, toys and park benches.

Recycling starts at home. You rinse food containers, crush cans flat and tear labels off cardboard boxes. That makes it easier for machines to separate each material. At the recycling plant, conveyor belts carry items past workers and robots that pick out anything that does not belong. Then each pile gets turned back into tiny pieces. Flakes of plastic, shards of glass, sheets of steel and pulped paper.

Manufacturers use those recycled bits to make fresh products. A plastic jug might become a cozy fleece jacket. Old newspapers turn into new newspaper rolls. Leftover aluminum from soda cans spins back into new cans in just a few weeks. Recycling one aluminum can conserve sufficient energy to power a TV for three hours!

By recycling, we help conserve natural resources. Fewer trees get cut down for paper. Less oil gets used to make plastic. Mining for metals slows down, protecting ecosystems where plants and animals live. Best of all, recycling keeps piles of trash out of landfills where waste can release harmful gases.

Challenge: Start a mini-recycling project at home or school. Label bins for paper, plastic and metal. Track how many items you collect each week. Then celebrate by crafting something new from your recycled treasures!

Green Roofs Help Cool Buildings

Have you ever seen a house with grass, flowers and shrubs growing on its rooftop? That's a green roof, and it makes buildings cooler and happier places. Instead of hot tiles that soak up sunshine and blast heat into rooms below, green roofs use soil and plants to absorb sunlight and release moisture.

During summer days, plants on the roof capture rainwater and let it slowly evaporate. This natural process cools the air above the building. Inside, rooms stay fresher without overworking air conditioners. That means families save on energy bills while plants enjoy a sunbathed home in the sky.

Green roofs also help cities stay cool. On sunny afternoons, concrete and metal reflect heat back to the air, making a city feel like an oven. Plants break that cycle. Rooftop gardens act like big green sponges that trap heat and keep streets below more comfortable. Birds and insects find new habitats high above the ground, too.

Installing a green roof involves layers of protection: waterproof sheeting, drainage panels, soil and then plants. Lightweight sedum, a succulent plant, often covers green roofs because it survives drought and stays low. Some rooftops grow tiny vegetables, lettuce and herbs that people can harvest for fresh salads.

Try This: Create a small green roof model on a shoebox. Layer plastic wrap, a sponge, soil and seeds. Watch sprouts grow, then test how that roof stays cooler than a bare shoebox under a lamp.

Water Filters Keep Rivers Clean

Clean water means healthy people and happy wildlife. Water filters help protect rivers by catching dirt, chemicals and tiny plastics before water flows back into streams. Filters come in many shapes. Some sit beneath kitchen sinks to purify drinking water. Others live inside big buildings to clean stormwater that washes off roads and rooftops.

In cities, rainstorms send polluted water rushing toward rivers. That water picks up oil slicks from parking lots, dust from roads and litter from sidewalks. Filters (made of sand, gravel and special membranes) act like giant coffee filters. They trap pollutants while letting clean water flow through into local streams.

On a smaller scale, simple charcoal filters can sit inside a jar. Pour muddy river water through layered sand and tiny gravel, then let it drip through charcoal. Charcoal captures bad smells and chemicals. The result looks much clearer, though you still need to boil it for safety.

Engineers design large filter systems with plants, rocks and wetlands in parks. These living filters attract frogs, birds and butterflies while cleaning water. Roots trap harmful sediments and microbes. Tiny water bugs munch on leftover nutrients so rivers stay healthy for fish and people.

Challenge: Build a mini-filter at home with a clear bottle, gravel, sand and activated charcoal. Pour muddy water through it and watch contaminants get trapped. Discuss why filtering slows pollution and why clean rivers matter for wildlife and communities.

Solar-Powered Gadgets Are on the Rise

Imagine charging your devices using only sunshine. That's exactly what solar-powered gadgets do: they carry tiny solar panels that soak up light. You might spot backpacks with panels stitched into their fabric. While you walk to school, they gather energy to refill your phone or power a small lamp at night.

Handheld solar chargers come in pocket sizes. Set one in the sun for a few hours, then plug in your device for a quick boost. Some flashlights store solar energy, so they never need batteries. Take one camping and it will shine all evening after a sunny afternoon hike. Even calculators use solar cells under their displays.

Garden sprinklers, Bluetooth speakers and outdoor lanterns also run on solar power. Engineers design each gadget to hold energy safely in built-in batteries. That way, your speaker plays music under starry skies and your lantern glows softly after dusk.

Solar gadgets work best under direct sunlight. In cloudy weather, they collect less energy but still keep trickling power into their batteries. A short time in the sun can mean longer use when clouds roll in.

Challenge: With an adult's help, build a simple solar charger using a small solar panel, a rechargeable battery and wires. Place it in sunlight to charge, then measure how many minutes of device use you get from 10 minutes of sun. Seeing sunlight turn into power makes solar magic real!

Bicycles Are Super Eco-Friendly

Riding a bicycle is one of the easiest ways to help the planet. Your feet turn pedals, the chain spins the wheels and you zoom down paths without burning fuel or making pollution. Every time you choose a bike instead of a car for short trips, you cut greenhouse gas emissions and reduce noise in busy streets.

Bicycles need little energy beyond your own legs and a bit of food for fuel. They require fewer resources to make than cars. A simple frame, two wheels and a chain can last for years with basic upkeep like oiling the chain and checking tire pressure. That low maintenance means fewer parts end up in landfills.

Many cities now build protected bike lanes so cyclists stay safe and keep traffic moving smoothly. Schools host "bike trains" where groups of students ride together with adult volunteers. Neighbors organize repair workshops to teach tire patching and gear adjustments. Riding a bike brings friends together while caring for Earth.

Cyclists help plants and wildlife too. With fewer cars on roads, urban green spaces thrive. Birds and insects sing without engine noise. Cleaner air gives everyone healthier lungs.

Challenge: Plan a "Bike Day" at home. Map a round-trip route under 5 miles, pack a recycled-paper lunch and invite family or friends. Track how much carbon you save by biking instead of driving.

Composting Gives Food Scraps New Life

Instead of tossing apple cores and banana peels into the trash, you can start a compost pile that turns food scraps into garden gold. Composting is nature's way of recycling. Bacteria, worms and tiny bugs break down kitchen leftovers and yard trimmings into rich soil that helps plants grow strong.

To compost, choose a bin with holes for air. Add equal parts green material, like fruit peels, vegetable scraps and coffee grounds and brown material, such as dry leaves, shredded paper and small twigs. Keep the pile moist, just like a wrung-out sponge. Every week, turn the mix with a shovel so air reaches each piece and speeds up breakdown.

In a few months, the pile shrinks into dark, earthy compost filled with nutrients. Gardeners mix it into flower beds and vegetable patches to improve soil structure, hold moisture and feed healthy roots. Also, houseplants gain new life when potting soil contains a bit of compost.

Composting also reduces landfill waste. Food scraps in landfills produce methane, a potent greenhouse gas. Composting at home or school means you help cut that pollution and shorten garbage pickups.

Challenge: Start a mini compost in a clear jar. Layer shredded newspaper, green scraps and a handful of soil. Mist with water, seal with a lid that has holes for air and place in a warm spot. Observe the layers each week and note how quickly they transform into rich soil. That tiny jar shows nature at work!

Eco-Houses Are Built to Save Energy

An eco-house blends smart design and green materials to use less energy and protect our planet. These homes face the sun in winter to capture warmth through big windows. In summer, overhangs block harsh rays, so interiors stay cooler without running air conditioners constantly.

Walls packed with thick insulation trap heat during cold months and keep heat out on warm days. Roofs often sport solar panels that generate electricity for lights, appliances and water heaters. Some eco-houses use small wind turbines mounted on their roofs to add clean power.

Rainwater harvesting systems collect water from rain in barrels. That water drips through filters, then waters gardens and flushes toilets. Greywater from sinks and showers goes through simple filters to reuse in irrigation. This lowers water bills and eases pressure on city systems.

Natural materials play a big role, too. Bamboo floors, recycled steel beams and low-VOC paints keep indoor air fresh. Green roofs covered in plants help regulate temperature, absorb rainwater and create habitat for birds.

Many eco-houses feature smart energy controllers that learn residents' daily routines. When the light and heat are adjusted automatically, these systems cut waste without anyone lifting a finger.

Challenge: Sketch your dream eco-house. Label where solar panels sit, how rainwater flows, and which materials protect the indoor air.

Ocean Clean-Up Robots Collect Trash

Our oceans face a big problem: plastic and trash drifting in currents, harming fish, sea turtles and birds. Ocean clean-up robots help tackle this challenge. These smart machines range from floating booms that guide debris into nets, to small underwater drones that hunt for microplastics.

One ocean robot is a solar-powered craft that glides across the sea, gathering floating rubbish with a giant scoop. It moves slowly so sea creatures can swim safely underneath. Once its hold is full, a support ship arrives and empties the trash for recycling back on land.

Underwater drones use cameras and sensors to spot plastic bags, bottles and fishing nets below the surface. Engineers program them to dive down, grab debris with a robotic arm, then rise back up. These submarines explore tight spaces around reefs where bigger boats cannot reach, keeping corals safe from damage.

Robots also patrol beaches, using conveyor belts to pick up cigarette butts, bottle caps and wrappers. They sort items by material (plastic, metal or glass) so each type heads to the right recycling plant. This makes sea trash useful again instead of letting it drift back into waves.

Ocean clean-up robots work best when people support them by reducing single-use plastics, recycling properly and joining beach clean-ups. Together with these machines, we can protect marine life and keep beaches beautiful.

Tech Word: Autonomous is a robot that can perform tasks on its own, without being controlled directly by a person.

Plant-Based Plastics Help the Planet

Traditional plastics come from oil, a nonrenewable resource that takes millions of years to form. Plant-based plastics offer a greener choice. Made from corn, sugarcane or potato starch, they behave like regular plastic but break down faster in the right conditions.

First, plants like corn grow in fields, soaking up sunlight and carbon dioxide as they grow. Farmers harvest the corn, crush kernels to extract sugars, then ferment those sugars into lactic acid. Scientists turn lactic acid into a plastic called polylactic acid (PLA). PLA looks and feels like a plastic called polyethylene terephthalate (PET) used in soda bottles, but it composts in industrial facilities within months instead of lasting centuries.

Plant-based plastics power packaging, cups, food containers and even toothbrushes. Some biodegradable bags use a mix of plant polymers so they are strong enough for groceries yet break down in a home compost bin. Other bioplastics combine plant materials with recycled PET to boost strength and eco-benefits.

These plastics still need proper disposal. If they end up in regular recycling bins, they can contaminate the recycling stream. Special composting programs help plant-based plastics return to soil, enriching gardens rather than piling up in landfills.

By choosing products made from plants, families support farms that capture carbon and reduce reliance on fossil fuels. Scientists continue developing new bioplastics from algae, mushroom roots and food waste, making the future of plastics even greener.

Try This: Gather clean plant scraps, banana peels, carrot tops and coffee grounds. With an adult's help, mix them into homemade paper sheets (blended with water and pressed flat). That "bioplastic paper" shows how plant fibers become new materials!

Kids Can Be Green Inventors Too!

You don't have to wait until you grow up to help the planet. Kids can be green inventors right now! All you need is curiosity, a bit of creativity and willingness to test new ideas.

Start by observing everyday challenges: Does your family throw away single-use sandwich bags? Can a reusable snack pouch be made from old fabric? Notice leaky faucets. Can a simple valve save gallons of water? Spot plastic pollution. Can a pocket-sized filter catch microplastics in puddles?

Next, sketch your idea on paper and list materials you might need. Gather recycled items such as cardboard, bottle caps, plastic containers and basic tools under adult supervision. Use hot glue, tape or simple screws to assemble your prototype. Don't worry if your first model looks rough; inventors learn by tweaking and testing.

Test your invention and take notes. Does your snack pouch seal tightly? Does your water saver valve reduce drips? Record what works and what needs improvement. Share your results with friends and family and maybe they have ideas to make it even better.

Enter a young inventor contest or host a mini science fair at home. Display your invention, explain how it helps the planet and invite feedback. Seeing others inspired by your work can spark new green solutions in your community.

Many famous inventions began as a child's simple idea. With persistence and passion, you can grow your prototype into a real product someday. Every little invention adds up to big change for Earth!

Tech Word: **Prototype** is the first simple version of an invention used to test ideas and make improvements.

Chapter 8:
The Future of Technology

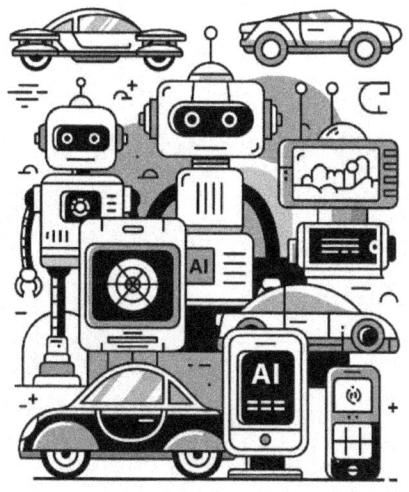

Flying Cars Are Being Tested Today

Imagine a vehicle that drives down the street like a car then lifts into the sky like a small airplane. That's the promise of flying cars and engineers around the world are testing them right now. These vehicles use electric propellers or small jet engines tucked into their wings. When traffic slows, the propellers spin up and the car rises above the gridlock. In tests, pilots guide flying cars along fixed routes like air taxis, carrying passengers between city rooftops.

Designers build sturdy frames from lightweight materials such as carbon fiber. Batteries store power for short flights of thirty minutes to an hour. Advanced software keeps flying cars level, avoids obstacles and helps land on designated pads. Some prototypes fold their wings and propellers so they can park in regular garages or drive on roads like today's vans.

Safety systems include parachutes that deploy if an engine fails and sensors that spot other aircraft in busy air corridors. Air traffic controllers will eventually manage lanes in the sky much like they do on runways. Companies aim to win government approvals for passenger flights within the next few years.

Flying cars could cut commuting time by half, connect remote towns and open new ways to deliver goods quickly. As batteries become lighter and charging stations become more common, these sky cars may transform travel for families, emergency medical teams and delivery services. A world where your next school run takes flight is closer than you think!

Robots Might Deliver Your Pizza

Picture a small, wheeled robot rolling down the sidewalk with a hot pizza in its insulated box. That scene is already real in some cities. Delivery robots use cameras, GPS receivers and radar sensors to navigate sidewalks and crosswalks safely. They map the route to your door, slow down for pedestrians and wait at curbs until traffic clears.

Inside, the robot keeps your pizza warm on a heated tray. When it arrives, it sends a notification to your phone. You open a locked hatch using a code, grab your slice and watch the robot roll away to its next delivery. Engineers build these machines from weatherproof materials, so rain, snow and heat do not stop them.

Robots learn from each trip. Machine learning software helps them handle new obstacles such as a stray shopping cart, a pigeon swooping down or an uneven sidewalk crack. If they get stuck, operators guide them remotely to free the wheels.

These delivery robots reduce traffic congestion and cut emissions by replacing cars or scooters on short trips. They work day and night, making late-night pizza runs easy. Soon robots may deliver groceries, medicine and small packages right to your porch. A dinner delivered by a whirring robot could become your new normal.

3D Printers Can Build Houses

What if a giant printer could build your home? That idea is coming true with 3D construction printers. These massive machines use layers of special concrete mixed with polymers. Controlled by computer designs, the printer arm moves side to side, adding one layer at a time until walls rise from the ground.

Unlike traditional construction, 3D printing cuts waste. The printer uses just the right amount of material, so extra concrete does not pile up. Building panels, curved walls and intricate archways appear without needing molds or templates. Workers guide the printer, install windows and doors, then add roofs once the main structure is ready.

Some startup companies have printed homes in 24 hours, complete with living rooms, kitchens and bathrooms. These houses meet building codes and come in different sizes, from tiny studios to family homes. Costs drop because labor needs to shrink; a single printer arm handles much of the work.

3D-printed houses help address housing shortages around the world. In disaster zones, printers can build shelters quickly after earthquakes or storms. Remote communities without local builders gain access to safe housing built by autonomous machines. Future printers may use recycled plastic or local soil to make walls, reducing transport costs and environmental impact.

A printer that builds houses could usher in a new era of fast, eco-friendly construction. Your next home might rise layer by layer from a digital file!

Holograms Will Make Screens Float in Air

Remember holograms from sci-fi movies? Soon, you may swipe, tap and watch them in midair. Holographic displays project images that hover without a screen. They use lasers or arrays of tiny LEDs to bounce light off special glass or mist, creating three-dimensional pictures you can walk around.

Early holographic devices use spinning mirrors to paint images so fast your eye sees a solid shape. Others shape beams of light directly, focusing them at precise points in space. Wearable headsets also deliver holograms by sending separate images to each eye, making objects appear to float in front of you.

Holograms could replace flat screens in phones, tablets and TVs. Imagine video calls where friends stand beside you as life-sized projections. Scientists could train surgeons with floating 3D models of organs, helping them practice tricky operations. In classrooms, teachers could show planets, dinosaurs or molecules in full 3D, sparking curiosity and deep learning.

Holographic billboards may shine from city centers, displaying moving art without bulky frames. Retail stores could showcase products in midair, letting customers inspect items from every angle. As light engines get smaller and more efficient, holograms will move from labs into homes, schools and shops.

A future where images float freely could change how we learn, play and connect, blurring the line between virtual and real in the most magical way.

Smart Glasses Could Replace Phones

Imagine slipping on a pair of lightweight glasses that do everything your phone can do. Yes! These are smart glasses. They pack tiny screens into their lenses. When you look at a friend, a message pops up in the corner of your view without hiding the real world. These glasses can show maps that guide you step by step, display alerts about your homework schedule and even translate spoken words into captions right before your eyes.

Built-in cameras capture photos and videos hands-free.

Want to share your soccer practice highlights? Just tap the frame or speak a command and your glasses record every kick and goal. Voice assistants live in the frames too, so that you can ask questions out loud. "How tall is Mount Everest?" brings the answer up and you can read it while you keep walking.

Some smart-glass models let you make and answer phone calls with a simple nod. Tiny speakers near your ears deliver sound privately, so only you hear your friend's voice. Because the electronics sit in the temples like regular eyeglasses, weight and bulk stay low. Battery life stretches through a school day and charging docks refill them quickly between classes.

Augmented-reality apps turn your world into an interactive game. Point your glasses at a tree and a virtual squirrel might pop out, chattering facts about woodland habitats. In art class, 3D shapes hover on your desk as you sketch. Science experiments come alive when virtual molecules float in the air around you.

Smart glasses could free your hands from screens, letting you learn, share and play without holding a device. Soon, these high-tech specs may become as common as smartphones, only sleeker and more magical.

Your Clothes Might Charge Your Devices

What if your shirt could power up your tablet? Energy-harvesting fabrics blend everyday clothes with tiny electronics that capture energy from movement, sunlight and your body heat. Scientists weave conductive threads into cotton, polyester or nylon. When you walk, the bending and stretching of these threads generate small electric currents. That power travels through hidden wires to pockets where you plug in a phone or a wearable.

Other smart garments use flexible solar panels sewn into jackets, backpacks and caps. During a sunny lunch break, panels soak up rays and fill thin batteries inside your sleeve. By afternoon, you might have enough juice to recharge earbuds while you study, no wall outlet needed. In fact, on cloudy days, energy trickles in thanks to sensitive photovoltaic fibers.

In winter sports, heated ski gloves use body-heat sensors to direct warmth where fingers get cold. Those same sensors can flip into power-capture mode when you stop for cocoa, sending extra warmth into fabric batteries instead. Designs stay comfortable and washable, so you can toss gear into the laundry after a muddy soccer match.

Textile researchers also explore piezoelectric materials, which are substances that produce electricity under pressure. Embed those into shoe insoles and every step you take generates tiny bursts of energy. Over a day of walking, your jacket's hidden batteries collect enough to light a safety blinker or power a GPS tracker on a family hike.

Clothing that charges devices brings energy right to your body. No more hunting for outlets, no more tangled chargers. As smart fabrics improve, you might head out in the morning wearing a fully charged power station and your own jacket.

AI Will Become Super Smart

Artificial intelligence (AI) is growing more powerful every day. Soon, AI systems will learn from far bigger collections of data, spotting patterns that humans cannot see. Imagine a virtual tutor that studies your math homework, sees which problems slow you down and then crafts new practice puzzles just for you. That AI adapts in real-time, making lessons as fun as a favorite game.

In medicine, super-smart AI can scan thousands of X-rays in seconds, finding tiny signs of illness that doctors might miss. These systems train on millions of medical images, learning to spot early warning signals for conditions like lung nodules or broken bones. That helps doctors treat patients faster and more accurately.

AI assistants will also handle everyday chores. They'll schedule your study time, order groceries when snacks run low and suggest healthy recipes based on what's in your fridge. Connected to home devices, AI could warm up the oven before you finish a cooking show or adjust lights and music for a cozy reading nook.

Robots powered by advanced AI will tackle complex tasks. In disaster zones, they'll search collapsed buildings using cameras and sensors, guiding rescue teams to survivors. In classrooms, friendly robot companions will help students who need extra support, speak dozens of languages and explain subjects in simple words.

As AI grows super smart, ethical design keeps it safe and fair. Engineers build guardrails so AI respects privacy and avoids bias. Governments work with tech teams to set rules that keep AI tools transparent so we understand how they make choices.

Super-smart AI promises a future where learning, caring and creating become easier, safer and more personalized.

Underwater Robots Will Explore Deep Oceans

The deep ocean remains one of Earth's last frontiers. Underwater robots are also called autonomous underwater vehicles (AUVs). They are built to dive where humans cannot reach. These sleek submarines carry cameras, sonar sensors and robotic arms. They dive through pitch-black water under crushing pressure to map the seafloor, study hydrothermal vents and search for new species.

Some AUVs resemble fish, using flexible bodies and fins to glide silently. Others look like torpedoes, shooting forward on propellers. Both styles collect data on water temperature, salinity and currents. Sonar beams bounce off rocks and gullies to create detailed 3D maps of canyons and underwater mountains.

When a robot discovers a coral reef, it uses high-resolution cameras to photograph fragile corals and sponges. Those images help marine scientists track reef health and spot bleaching events. Robotic arms gently pick up tiny samples of sediment without disturbing the environment, carrying them back to surface ships for analysis.

Deep-sea AUVs also inspect pipelines, cables and offshore structures. They check for leaks and corrosion in places too dangerous for divers. That work keeps energy companies and environmental teams aware of potential hazards before leaks spread.

Powered by long-lasting batteries or tethered to ships, these robots can stay underwater for days and weeks! They follow pre-programmed routes, avoiding obstacles using sonar "eyes." When their mission ends, they surface and send data to scientists in real time via satellite link.

Underwater robots bring the hidden ocean into view, revealing creatures and landscapes we've never seen. They guide efforts to protect fragile habitats, inspiring kids to explore the blue world beneath our waves.

Space Travel May Become Common

You may be too young to remember, but at one time, going to the moon felt like a fairy tale. But soon, it could be as normal as a plane ride. Companies are designing rockets that launch tourists into space for a few minutes of weightlessness. This is enough time to float like an astronaut and see Earth's blue curve beneath you. Space hotels are in the works, too: imagine drifting between rooms with walls made of windows showing stars all around.

Regular trips to low Earth orbit will let scientists run experiments in zero gravity, growing new materials that can't form on the ground. Kids might send artwork up to hang in space galleries. Even school classes could join virtual field trips to orbit, watching astronauts water floating plants before they orbit back for a live Q&A.

Strap in for quick suborbital flights that reach the edge of space, then glide back to Earth in just a couple of hours. No spacesuit needed; just a comfy cabin and exciting views. As rocket costs fall, daily launches could carry people and small cargo to space stations, lunar bases or Mars way-stations.

One day, packing for summer vacation might include flying thirty kilometers straight up, seeing sunrise every ninety minutes, then coming home with amazing photos. Proof that space travel isn't for history books anymore.

Smart Homes Will Do Everything for You

Imagine waking up and having your home-brew coffee, opening the blinds, and adjusting the room temperature all before you open your eyes. That's the future promise of smart homes. Rooms filled with sensors learn when you arrive, brighten lights on cloudy mornings, and play your favorite tunes as you walk through the door.

Refrigerators will track what you eat, suggest recipes based on your groceries, and even order milk when you run low. Floors embedded with pressure sensors can tell if you've fallen, sending alerts to family members. Windows automatically tint to block harsh sunlight, then clear up again when you need warmth.

Robotic helpers will handle chores: vacuuming corners, folding laundry and watering plants. Walls might display family photos or live video from backyard cameras, so you never miss a pet's antics. Bathrooms will know your morning routine, pour just the right amount of soap and adjusting the water temperature exactly how you like it.

Security systems won't just lock doors; they'll recognize friendly faces at the door and unlock for you. Voice assistants will understand questions in any tone, asking "What's the weather?" will bring up simple, spoken answers on kitchen speakers or even projected in midair above the countertop.

Soon, homes will feel like caring partners, making daily life smoother, safer and a little more like magic.

Virtual Reality Will Feel More Real

Put on a VR headset today and you will step inside another world. Tomorrow, that world will feel so real you'll forget you're wearing goggles. Haptic suits give gentle taps and vibrations, so when you touch a virtual cat's fur or swing a sword, your body feels every nudge. Special gloves let you feel shapes in midair!

Smell generators add scents: walking through a digital forest brings the fresh pine aroma, while a virtual bakery wafts warm bread your way. Wind machines and temperature controls create breezes and cozy warmth to match the scene. Best of all, wireless full-body trackers erase cables, letting you dodge, dance and leap without getting tangled.

Social VR spaces will let you meet friends' avatars at the beach, share a roller-coaster ride across continents or study in a floating island classroom. Teachers will guide history lessons by recreating ancient cities around you while scientists model molecules at human size so you can walk right through DNA strands.

As graphics improve and gear shrinks, VR will blend seamlessly with reality. Full-sensory simulations could train astronauts inside Martian habitats or help you practice public speaking before a friendly virtual audience. The future of VR is stepping into your wildest dreams and feeling every bit of them.

Self-Driving Cars Will Take You Places

Picture your family hopping into a car that drives itself. You buckle up, tell it, "Take me to soccer practice," and seats slide into expert mode. No hands on the wheel, just conversation and laughter. Autonomous vehicles use cameras, radar and lasers to read traffic lights, spot pedestrians and plan safe routes around roadworks.

Inside, entertainment systems adapt to your mood: homework apps switch on for study time, and music playlists update automatically based on your schedule. The car's AI monitors road conditions (slowing down for rain or steering around potholes) while sending real-time data to other cars so they work together like a well-coordinated dance troupe.

Self-driving buses will patrol neighborhoods on demand, stopping anywhere along a route. Robo-taxis could pick you up from home, drop you at a friend's house, then zoom off to help someone else. Delivery vans without drivers will bring packages to your door anytime, even late at night, without a single honk.

Specialized shuttles will serve airports and theme parks, guiding visitors while giving narrated tours. Wheelchair-accessible vehicles offer newfound freedom for people with mobility challenges, opening up adventures that used to require help from a driver.

As laws adapt and public trust grows, self-driving cars may become the norm. This will turn commute time into fun, safe and productive moments on the move.

Food Will Be Grown in Vertical Farms

Imagine wandering through a skyscraper filled with rows of lettuce, strawberries and tomatoes stacked one above another like colorful shelves. That's a vertical farm. Instead of spreading across fields, plants grow in tall towers under special LED lights that mimic sunshine. Water drips down from the top, nourishing every leaf without wasting a single drop. Sensors measure moisture, temperature and nutrient levels, so each plant gets precisely what it requires to grow strong and tasty.

Vertical farms can sprout fresh greens right inside cities. Farmers set up these farms in old warehouses or shipping containers, cutting out long truck rides that send fruits and veggies on marathon journeys. That means lettuce harvested in the morning could be on your salad plate that afternoon crisp and full of flavor.

Because everything happens indoors, pests and bad weather don't ruin the crops. No spraying of harmful chemicals; only clean water and gentle light feed the plants. Farmers program robots to clip ripe leaves, plant new seeds and check each tower for growth. This high-tech garden runs day and night, feeding families in neighborhoods that once had few grocery choices.

Vertical farms show how skyscrapers of green can feed our future. Picture a world where every city block grows its own dinner, which is fresh, healthy and right outside your window!

Your Toys Might Talk Back

Have you ever wished your teddy bear could answer questions or your toy car could tell you when it's time to zoom? Soon, playthings will come alive with talking toys powered by tiny computers and clever sensors. Inside each toy, a small microphone listens when you speak. Your words travel through a mini "brain" (an AI chip) that decides how to respond. Then a speaker in the toy's belly shares its answer in a friendly voice.

These chatty toys learn your name, remember your favorite games and even tell jokes when you need a laugh. In the dollhouse, tiny robots act as family members, inviting you to tea parties or detective missions. Plush animals share bedtime stories in soft whispers, helping you drift off to sleep. Race cars announce lap times and challenge you to beat your best speed.

Smart toys also teach as they play. A talking globe explains where each country lies, while a building-block robot shows you how to stack towers without toppling them. Language toys guide you through new words in Spanish, Mandarin or sign language, turning playtime into an adventure around the world.

With safety in mind, these toys use kid-friendly filters, so they always give helpful, age-appropriate answers. Thanks to talking toys, playtime becomes a two-way conversation, sparking curiosity and creativity with every question you ask.

Kids Will Invent Tomorrow's Tech

You might think inventing world-changing gadgets is only for grown-ups, but history tells a different story. Many great breakthroughs started as simple ideas from kids like you. In the future, children will create the next generation of tomorrow's tech, from eco-friendly robots to pocket-sized labs that test water quality with a drop.

Imagine a 10-year-old designing a wristband that turns hand motions into music so you can compose symphonies in midair. Or a group of friends building a drone that plants trees where forests have burned, all controlled by an app they coded themselves. In colorful maker spaces, you'll find soldering irons, 3D printers and laser cutters. Tools that bring sketches from spiral notebooks into real prototypes.

School projects will involve real engineers who mentor students through challenges. Science fairs transform into invention expos where judges include top inventors and entrepreneurs. Winning ideas earns lab time and funding to grow a simple model into a life-changing product.

Remember, every technology begins with asking, "What if...?" So grab a pencil, sketch your wildest idea and test it with bits of cardboard, string and spare circuits. Share your prototypes with friends and refine them together. With imagination and a willingness to tinker, you could be the inventor whose name headlines tomorrow's tech news because you changed the world with one brilliant idea!

About The Author

Marcus Morales

Marcus is passionate about helping children learn about STEM (Science, Technology, Engineering and Mathematics) industries. A tech enthusiast himself, he grew up inspired by technology and the potential for a future where humanity solves our greatest problems. He works diligently with other staff at Intel-Excellence to bring tech related educational material to children throughout the world.

For more information visit www.intel-excellence.com

www.ingramcontent.com/pod-product-compliance
Lightning Source LLC
Chambersburg PA
CBHW031644040426
42453CB00006B/209